MOMENTUM

MOMENTUM

SIX PRINCIPLES PRODUCT
LEADERS FOLLOW TO **ENGINEER**
GOOD PRODUCTS FASTER

Shiva Nadarajah *and* Suresh Kandula

HANSEN PUBLISHING

H

HAMDEN PUBLISHING

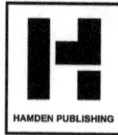

MOMENTUM
SIX PRINCIPLES PRODUCT LEADERS FOLLOW TO
ENGINEER GOOD PRODUCTS FASTER

ISBN 978-1-5445-2904-2 Hardcover
978-1-5445-2903-5 Paperback
978-1-5445-2902-8 Ebook

This book is dedicated to:

My wife, Sarah,
and my sons, Jack and Grant
—Shiva

My wife, Jayanthi,
and my sons, Abhinav and Charan
—Suresh

And to all those who inspired us over
the decades.

CONTENTS

INTRODUCTION

"Ideas are easy. It's the execution that's hard."
—JEFF BEZOS

Imagine a product manager, an architect, and an engineering team lead—collectively, a product leadership team—have gathered informally to debrief and vent their frustrations about recent events and the direction in which their organization is heading. The product manager walks through feedback from the latest customer survey, which indicates dissatisfaction with a new product, and is dismayed to report that a recent news article has highlighted a competitor's success with a similar product. The architect details the Herculean effort put forth by the production team to roll out a new feature—only to receive similarly disappointing results, certainly disproportionate to the additional strain on the team. The engineering team lead, who is new to the organization and who, therefore, has the benefit of relative objectivity, states that he finds it astonishing that there was such a delay and degree of difficulty in releasing a capability into production, given the extent of the enterprise's considerable resources.

They argue in circles as to the cause—is it the legacy architecture, an ancient mainframe, outdated modes of delivery or an incorrect code delivery approach, lack of automation, a waterfall methodology instead of Agile, or simply having teams set up inefficiently? Whatever the cause (or causes), they can all agree on one conclusion: if this pace continues into future quarters, their organization is not going to survive in the marketplace. Startups are faster, and digital natives are able to innovate and deliver more quickly, but large enterprises are generally unable to keep pace. Customers have noticed that other more cutting-edge solutions exist elsewhere—and the internal teams have also noted that their organization is behind the innovation curve.

They know they need to find a better way to deliver a better customer experience. They know they need better architecture. They know they need more momentum, but coming to this solution only presents another problem—one which leaves the product leadership team paralyzed—and that is the problem of not knowing where to begin.

MOMENTUM

No matter the specific role, any member of a product leadership team can recognize when their organization is not having the impact in the marketplace it should be having. The symptoms of that problem may include seeing competitors or new entrants in the market rolling out new capabilities and innovations faster than they are, being unable to meet customer expectations of types of products or flexibility within existing products, lack of growth compared to industry standards, and an exodus of both customers and internal team members. As satisfaction drops, there is a secondary impact in revenue, profit, and other metrics of the health of an organization. Talented employees are going to do what's best for their careers and make the move to a company where innovation happens much faster, leading to problems with attrition.

This becomes a multidimensional engineering and organizational challenge, and a product leadership team is unlikely to know where to begin or how to structure their change initiative to balance the dichotomy of being comprehensive but also expedient.

Within financial services, as an example, this challenge of lack of momentum is quite visible. According to a McKinsey study, on average only 30 percent of an organization's technology spend is directed toward new business capability-oriented development, whereas 70 percent is consumed by re-engineering, support activities, and infrastructure maintenance.[1] As a byproduct of this, smaller, nimbler competitors have flourished, with annual private equity funding growing 10X from roughly $4 billion in 2010 to $40 billion in 2020. Enterprises attempt to match this investment with their own spending; as an example, retail banking spent $20 billion on digital technology in 2017. Yet, dollar for dollar, large well-funded enterprises lag in getting new capabilities out to the market and are unable to match the momentum of their new digitally native competitors.

But what does momentum mean? Momentum is measured as the product of the mass and velocity of a moving body. In the context of business, momentum refers to the amount of impact an organization creates, along with the speed at which they create that impact. An organization is not going to focus solely on how fast they can create something; this is not about speed for speed's sake because misdirected hustle or quickly rolling out the wrong features is counterproductive. Instead, momentum acknowledges the importance of speed, but it must be combined with impact—doing business the *right* way, holistically in terms of both the customer and the organization, with the right architecture, right operational metrics, and right principles. Momentum is a powerful force in any organization. When leaders direct it correctly, they can start to see positive feedback that drives their momentum even higher!

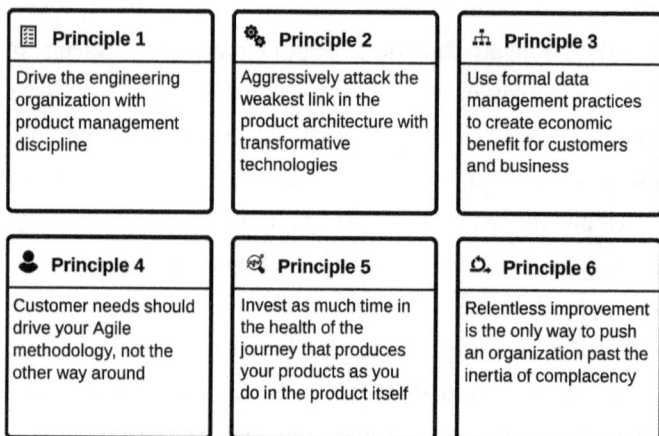

Principle 1	Principle 2	Principle 3
Drive the engineering organization with product management discipline	Aggressively attack the weakest link in the product architecture with transformative technologies	Use formal data management practices to create economic benefit for customers and business
Principle 4	Principle 5	Principle 6
Customer needs should drive your Agile methodology, not the other way around	Invest as much time in the health of the journey that produces your products as you do in the product itself	Relentless improvement is the only way to push an organization past the inertia of complacency

Figure 1: Six Principles

OUR MOMENTUM METHODOLOGY

Through our experiences and research, distilled from interviews with executives, product managers, analysts, and engineers, we have identified the six principles that most impact a product team's momentum, aiding them on their transformation journey to evolve into agile, resilient, product-oriented organizations rooted in a customer-centric view. If the reader follows the principles outlined in this book, they will begin to make the step change necessary to get their teams on a path to be more customer-centric, make better engineering decisions about their business, modernize their architecture more efficiently, extract more value out of their data, and be able to better structure teams to drive more momentum in an agile way.

The reader will also learn:

- *How to infuse product management thinking throughout their teams (not just product owners)*
- *How to build an effective enterprise architecture roadmap*
- *How to extract the most value from their data*
- *How to evaluate the organizational structure and its effectiveness, and select the right framework that can unlock human potential*
- *How to build and manage a resilient technical delivery pipeline*
- *How to avoid stall and set their teams up to constantly evolve and grow*

These six principles are interconnected, and this book serves not only as information about the principles themselves, but also gives a method of how to approach them. This framework allows teams to see the broader picture and manage the interdependencies across the principles. Product leaders need framing in order to have holistic conversations with their teams; this framework provides a structured way to approach change discussions with the proper perspective.

Absent this framework, it is all too common for teams to find themselves mired in myopic conversations, focused on a particular issue absent of the larger context of that product team's momentum. For example, triggered by a vendor mandate to cease supporting a particular off-the-shelf product, a team may go down the rabbit hole of reacting to that urgent risk and pursuing a swap option in isolation. When perhaps, with the benefit of a framework, the team may either have had a proactive strategy in place or could route a path forward in the context of the product's broader data strategy.

For large financial services organizations, change tends to be reactive—a result of new regulatory requirements, burning issues, end of life technology, scalability risks, or slow delivery infrastructure. The principles allow a constant evaluation of these risks and introduce a methodical way of dealing with risk and change. With this framework in place, the team will be able to place issues in the context of the factors that drive product momentum and discuss approaches in relation to all six principle areas, always keeping focus on the impact to the business.

The interconnected nature of complex product delivery issues cannot be solved in isolation. A product leadership team trying to retire a mainframe also has organizational challenges to consider. They will have ambitions to integrate new technologies beyond moving away from the mainframe to enrich customer experience. They must be mindful of the business capabilities they're ultimately delivering, as well as what that trend line looks like. Finally, they need a sense of how reliable their new solution is going to be. Even the seemingly ring-fenced case of shutting off a mainframe shows how these principles are interconnected.

Leading with business needs acts as a forcing function to help cut through this web. Product leadership teams shouldn't attempt to solve the technology problem, the data problem, or the people problem; they should instead solve the *business* problem. With a top-down view of what's working and what's not, they can explain to their management, in business terms, the root cause of the problem to be solved, as well as the magnitude of the issue from the perspective of cost, reputational risk, and customer voice, all tied back to the business case. This framework is not just a method to madness in terms of managing or a way of solving one single problem; it's a way to have a discussion in order to determine the overall business problem that needs to be resolved.

This framework also enables the product leaders to assess their maturity across the principle areas. Maturity must be developed based on where the group is starting and what is relevant to the team's specific business capabilities. By pushing them to take a particular principle to the next level of maturity, the team's skillset improves and their mindset evolves, becoming broader and more strategic.

Product leadership and management should also be aware of the balance between business context and technology understanding so they can lead their team to success. Without a clear grasp of both sides, they run the risk of suboptimal product delivery velocity, which will inevitably create unrealistic expectations for engineers, as well as frustration from leadership who want something done on unrealistic timelines because too much time has elapsed since they received progress updates.

With this framework in place, however, the entire organization will be faster to market, more in touch with the voice of the customer, and better positioned to impact business performance and growth.

WHY WE DO WHAT WE DO

Over the past two decades of working together, we have had an opportunity to work with many large financial services firms. Being in consulting has allowed us to benefit from the fusion of business and technology perspectives. Leading consulting teams with interdisciplinary backgrounds allows us the opportunity to think strategically with our clients about our industry, but also to understand the impact technology can have to create new, innovative products.

We both have computer science and engineering backgrounds, have managed global product and engineering teams, and have worked with several large financial institutions. We have had the opportunity to work together with clients who manage portfolio product teams in a variety of circumstances. In our journey, we have often found that the challenges our clients face tend to anchor to one or more of these principles. We have been successful in helping our clients navigate the change, evolve their organizations, create better teams, and release better products—and that has been inspiring.

Financial services are the backbone of our society. We have a passion for this industry because of the opportunities it provides people to secure their future and help sustain future generations. This book is a small contribution to financial services product engineering teams who want to see their collective vision become a reality.

WHY WE WROTE THIS BOOK

We wrote this book because, as we've implemented the framework with our clients, many have said, "I wish I'd known about this when I was first getting started." We've had unique insight into what's worked for them—and what hasn't—in our view of partnering with Fortune 500 clients over more than two decades. We wanted to write a book that leverages those experiences, mistakes, and successes. We intend for those learnings to be distilled into actionable principles and methodology that can be applied in a practical setting with confidence.

We've also noted over the years that certain key deliverables were instrumental in guiding effective conversations with teams. We've captured these artifacts throughout the principles, referring to them as "on the wall." This is our way of giving a nod to the power of war rooms and physical artifacts. Their digital equivalents are necessary and can be as effective; however, when we've been in the trenches doing this work for product leadership teams, we have found war rooms and physical artifacts still reign supreme.

Now, more than ever, there is increased pressure on the product leadership team to demonstrate results faster. There is less patience for delay. At the same time, the competition for top force-multiplying talent has grown exponentially. Technology moves quickly, so organizations have to be able to react faster than they would have had the luxury of reacting in years past. Consumer expectations are always evolving, but organizations also internally expect more in terms of product behavior month over month rather than year over year. The internal focus of companies is moving from a technology, business, or operations focus to a more product-centric model, making it more efficient, agile, and nimble. With the Momentum principles, our goal is to empower product leaders so they not only keep pace with increasing demands, but proactively work to improve their organization's processes for increased speed and measured responsiveness.

WHAT THIS BOOK IS AND ISN'T

This book is a guide that allows product leadership teams to apply a holistic mental model to solving issues that impede team momentum. On each of these subjects, there are reference books that go into significantly more detail, and we do not intend this book to be a substitute for domain-specific deep dives. Instead, it is a guidebook that shows a path through the six principles we've chosen to focus on.

We also want to be clear that our book does not comprehensively cover every domain product leadership teams have to focus on, such as business strategy, sales and marketing, branding, or efficiently managing customer requests. Product leaders will play a role in these areas and be expected to have an educated opinion about them, but this book is not going to address them in depth.

We've found that velocity and momentum are influenced by these six principles. A production team rarely fails to reactquickly to the market because their user experience design process is slow or because their marketing function isn't calibrated right. Instead, these six principles have an outsize influence over having an efficient, agile, and quick reaction product organization.

The next six sections will cover each principle in detail, moving through the maturity model while discussing specific deliverables that should be on the wall, and ending with

A chief technology officer (CTO) newly joining a financial services company experienced this pressure and, unfortunately to his detriment, failed to implement all six principles. Within his first two months on the product leadership team, he attempted to boil the ocean by overhauling the organizational structure and completely rebuilding one of their core platforms—a full-court press on principles focused on organization structure and architecture.

Without a cohesive conversation with his team about what business capabilities he was trying to enable, the impact those changes would have on the overall operations of the firm, or even how he could rationalize the architecture, his presentation to the board did not go well. Ultimately, it was a waste of time and of his team cycles, he disappointed his team by not being able to make an impact, and he suffered reputational damage by going guns blazing into a board meeting without sufficient rationale or presenting a comprehensive point of view. Had he been thinking more holistically, he may have solicited feedback on all six principles, calibrating the appropriate step change on each.

Having this framework will prevent exactly that type of danger because it encourages product leadership teams to think of all six principles when they consider an important portfolio decision.

Consider, alternately, an insurance company with the objective of improving their loss ratio by 5 percent in the next five years. By progressing through each of the six principles, they were able to determine the best way to improve their business capabilities, highlight heat maps, use the data to figure out where money was being lost, and have a conversation with their team about what kind of processes they could improve in the next five years to make this loss ratio work—an improvement of some $170 million.

the key takeaways to memorialize the actionable steps. The first principle is about business and customer context because a product leadership team needs to be able to drive the engineering organization with a shared mission and goals.

PRINCIPLE 1

Drive the Engineering Organization with Good Product Management Discipline.

"Know yourself and you will win all battles."

—SUN TZU

Business, like war, follows a set of rules and techniques that must be mastered to succeed on the—virtual or literal—battlefield.

To achieve success, the modern army general imparts a few key understandings to his troops: First, they need to understand *why* they are fighting. What is the mission? What is at stake? What is the end goal, and how do they intend to achieve it? The general communicates his *strategy* to the troops.

Next, the general steps back to observe a full view of the landscape. The modern army general stands not on the front lines of action but instead sits in a command tent, far behind enemy lines, looking at the entire theater of war. From his more distant vantage point, the general sees the entire battle space holistically. That all-encompassing view allows him to evaluate the assets at his disposal and determine the total objective to be accomplished, the knowledge of which enables him to strategize the best plan of attack.

Finally, the general seeks to understand the end goal—both the immediate opposition they're fighting, and for whom those enemy troops are fighting: Is it the local population or a political figure or dictator? Who benefits if the enemy wins the battle or war? What are their needs and goals once the battle is won? General Stanley McChrystal, a former commander of US forces, in describing the goal of his mission, would say that his job was not just about bombing the enemy; he also had to provide guidance for how to support the local population after routing out the enemy.

Similar to a general artfully applying the rules of war, a product leadership team will be more successful at instilling good product management discipline if they are armed with knowledge of strategy, a comprehensive view of an organization's business capabilities, and a deep understanding of customer insights.

GOOD PRODUCT MANAGEMENT DISCIPLINE

Product management discipline means a strong connection between the product being built and the goals of the business. For example, if the organization's objective is to boost revenue, the teams should be aligned with building products or developing capabilities that will serve to increase revenue. If the priority is to improve margins, then the focus should be on enhancing products in such a way that operating costs will be reduced for the company, in alignment with the business strategy.

Historically, misdirected hustle occurs when teammates who do not share this knowledge decide to build something that has no direct business value. They may be driven to take advantage of the latest cutting-edge technology, which might also support their strategy or the wants and needs of the customer. This approach, however, is misguided. Instead, the team's focus should align with the business strategy and building capabilities that will support that strategy. Like a general approaching a new battlefield, a leader trying to increase momentum needs to assess their teams against this principle: *"How ready are my teams to take on this responsibility of having strong product management discipline?"*

To run a team with strong product management discipline, team leaders must adopt—and communicate—a mindset that incorporates a three-pronged approach:

1. **A CONNECTED BUSINESS STRATEGY:** Engineering aligned with product vision, mission, and business strategy means that, as the team prioritizes work, they will not have any misdirected hustle and will work toward developing products that support the firm's goals.

2. **COMPREHENSIVE BUSINESS CAPABILITIES:** Well-researched business requirements translate into a more impactful product. Additionally, a holistic view of an organization's capabilities allows product leadership teams to accurately map their business functions, which leads to better-informed, more effective decisions. These functions are categorized and prioritized in a business capability map that provides a bird's-eye view of the business function, as well as the importance of each of those pieces. With this all-encompassing perspective, the product leadership team can strategize the best course of action within the context of how it will affect the overall business.

3. **A CUSTOMER-CENTRIC BUSINESS ARCHITECTURE:** By injecting more customer-centric data into the decision-making process, the product leadership team is better informed to make decisions on product behavior, product feature choice, and feature prioritization. Whereas in the past, they may have simply relied on an internal team of experts or may have been swayed by senior leaders' opinions, establishing customer-centric journeys becomes a conduit to ensure that teams are building products with that customer in mind—not their internal opinions on what the customer might want.

All three of these areas are necessary for an engineering team to develop strong product management discipline. Focusing only on having a customer-centric point of view, for example, may lead to developing capabilities that don't generate the most revenue, or sustainable profit, for the firm. Because customers tend to say that they want many features available, the wants of the customer must be balanced with the business strategy—in terms of features being developed with robust business capabilities—and the potential revenue that will be generated.

The most successful product leadership teams we've seen are able to manage risk across their portfolio and settle on clear priorities for their team because the visions and goals they have match up to their framework. They can minimize misdirected hustle, collisions, or features going out that don't add value to the customer or that aren't a core differentiator for their organizations. With a clear business context, they can determine the best way to fix problems that arise—and prevent some problems in the first place, all with the benefit of perspective.

Like the general, product leaders should look at all the assets available on the battlefield to increase the team's chances of success.

UNDERSTANDING BUSINESS STRATEGY

As part of a good product management discipline, the team should be knowledgeable about the business strategy in the context of the portfolio of products the product leadership team is working on and how those products earn money or meet goals for the firm. Generally, most organizations have some semblance of a business strategy and a strategy by product; it is less common, however, for that knowledge to be disseminated across the entire team.

> Developing and imparting knowledge of each firm's business strategy to the product teams is a challenge of its own because it is firm specific.

Teams need to understand the basic value proposition of the company. Why would someone come to their firm to do business instead of going to some other company to do that same business? For example, a large bond broker assessing their competitive positioning against a digital retail trading upstart determined that their bond offering was five times larger than their competitor. Not only was the size of their pool of bonds used as a market differentiator, but in addition, the product team also built offerings that were able to take advantage of their breadth of inventory to showcase value to the customer.

The leadership team needs to percolate this knowledge down so that everybody knows specifically how the firm makes money. This key piece is often missing, so the product leaders should assess: does the entire product team understand the business strategy for this specific product portfolio? Can they answer the following questions?

- *What do we do?*
- *How do we do it?*
- *Why do customers come to us?*
- *How do we make money from those customers? What do they pay us for?*
- *What do we want to do in the future to make more money or grow this product?*

At the most rudimentary level, the entire team needs that knowledge of the forward-looking business strategy of the product.

At a more intermediate level, this aspect of the principle evolves such that the work the team does is connected either to the elements of the business strategy or to the specific indicators that the work is going to affect. At this stage, not only does the team understand the business strategy, but the product team then has a clear line of sight as to how these sets of features and epics they are working on are important because they also see exactly how the features apply to the strategy of the product before them.

Certain key performance indicators (KPIs) are used by the firm to measure the health, or success, of the product. For a stock brokerage firm, as an example, these KPIs may include the number of new customers, number of new accounts opened per customer, number of accounts being closed, the time it takes to open an account, registration abandonment rate, number of trades made per account, time to execute a trade, trade error, etc. The work that the team is doing is connected either to a specific business strategy or toward improving specific KPIs for the firm.

When employees have greater appreciation of the firm's business strategy, teams have the data they need to be able to make trade-off decisions. Teams are armed with information such as revenues and expenses associated with a feature over a reasonable time horizon, net effect on cash flow over that time horizon, and the net present value of the feature. Focusing on optimizing returns to the business allows teams to prioritize sprints objectively and think in terms of the business. Key assumptions are validated. After-action reviews serve as a feedback mechanism to continue to improve the team's accuracy in forecasting business success through product strategy.

These metrics are going to vary by business, but there are a few frameworks product leads can follow to drive these metrics, independent of the business they are in. These include:

AARRR FRAMEWORK: The AARRR framework, originally developed by venture capitalist Dave McClure, helps young startups grow. It is a five-step process that focuses on five key metrics: acquisition (in terms of the customer's first experience with the product), activation, retention, referral activity from existing customers to new prospects or those who came back to the organization for future purchases, and revenue generation.

KANO MODEL FRAMEWORK: When a product team has to deal with tough budgetary decisions, they rely on the Kano Model. The model can help them weigh high-satisfaction features against their costs of implementation so that they know which ones are worth adding and those that aren't.

HEART FRAMEWORK: Google created the HEART framework to bring quantitative metrics to a world of user experience that is typically ruled by qualitative data. The five measures (happiness, engagement, adoption, retention, and task success) help make decisions based on facts rather than feelings, with Google at the forefront of bringing this new form of analytics into product teams around the globe.

DEVELOPING COMPREHENSIVE BUSINESS CAPABILITIES:
Business Capabilities Map

A business capabilities map allows the product leadership team to visualize the organization's holistic capabilities, as in the army general example, by combining the outside-in view—from customer, partners, and regulatory points of view—with that

A Bain & Company survey of more than five hundred senior executives found that despite devoting enormous resources and energy trying to align IT investments with their most important business needs, fewer than one in five felt their efforts were succeeding.

Organizational alignment, as defined by *Harvard Business Review* authors Jonathan Trevor and Barry Varcoe, is when business strategy, business purpose, and capabilities are in-sync to execute on the same objective. Our view is that large organizations tend either to develop product capabilities with subpar customer experience or great customer experience via technology with subpar product capabilities. The ideal situation is to put the customer first and develop an aligned strategy for product and technology teams.

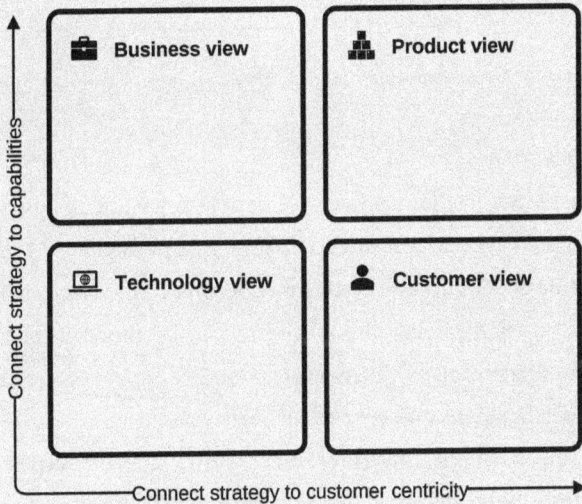

Figure 2: Aligning strategy to capabilities and customer needs

A CONCEPTUAL VIEW OF A BUSINESS CAPABILITIES MAP

Business Capabilities

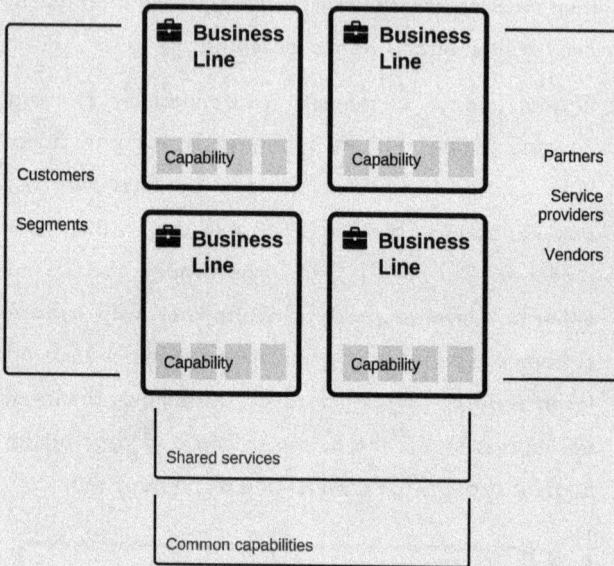

Figure 3: A conceptual business capability map

of inside-out, from the point of view of internal line of business and shared services. Business capability mapping depicts *what* a business does to reach its strategic objectives (its capabilities), rather than *how* the business does it (its business and technology processes). Business capabilities are the connection between business strategy and business execution.

The business capability map allows specific types of questions to be answered by product leadership teams.

- *What functions does the product or portfolio of products perform?*

Our approach is to build the capabilities map by working closely with subject matter experts in product and engineering organizations to expand on customer segmentation (example: Retail, SMB, International), specific line of business (example: Trading, Financial Advisor), capabilities (example: Stock Trading, Portfolio Planning), partners (example: Docusign, Salesforce), and shared capabilities (example: Marketing, Sales, Correspondence).

The accomplishments of the organization provide a clear view of how to approach an existing or, more importantly, a new problem statement.

- *What exactly do we do here? What is the detailed view of the product's functions?*
- *How efficiently do we do what we do? What are the processes and who are the people involved?*

Answering these questions through the lens of the business context will demonstrate areas of efficiency or inefficiency, potential pain points or issues, areas that generate revenue, or areas that may not need much focus because that capability doesn't make money for the company.

Product management discipline is the first principle in the

framework because the answers generated will create a common vocabulary that product leaders can use with their teams. In creating the necessary artifacts for this principle, particularly the business capability map, the product leadership team establishes the taxonomy, which can then be used to answer questions in various settings. They can ask their technology team, "How many people are working on client reporting?" with no confusion as to what that capability encompasses.

After creating a business capability map with the head of product at a large insurance company, we were able to identify some of the hot spots of their systemic problems to help improve their loss ratio, a key metric in the insurance industry. Over the years, this company had inorganically grown their claims processing systems and operations into seven different legacy claims platforms with different people, processes, and teams managing them across the country. Streamlining them into a single platform could improve the loss ratio by better managing their claims. Prior to creating these shared artifacts, it was difficult for the executive team to appreciate the complexities of managing the variety of insurance products across the multiple acquired companies that had been folded into the master company. A senior vice president walked up to the diagram on the wall, pointed to the claims capability, and, without our explanation, deciphered the root cause: the seven associated claims systems they had in use for different lines of business. He was then able to say, "This is the problem we need to solve."

The head of product hadn't initiated this process by stating, "We need a business context diagram so we can solve the problem of too many claims systems." The business had determined that a pain point existed, but the problem itself was unclear until this diagram went on the wall. Similarly, nothing had changed within the business itself, yet a simple, powerful document showed exactly what problem needed to be solved and how that solution would affect the rest of the business.

DEVELOPING A CUSTOMER-CENTRIC POINT OF VIEW:
Customer Journey Value Streams

A McKinsey analysis of ten years of growth demonstrated that firms with higher customer satisfaction grew in value four times relative to the laggards in their respective category.

Ten-year value growth

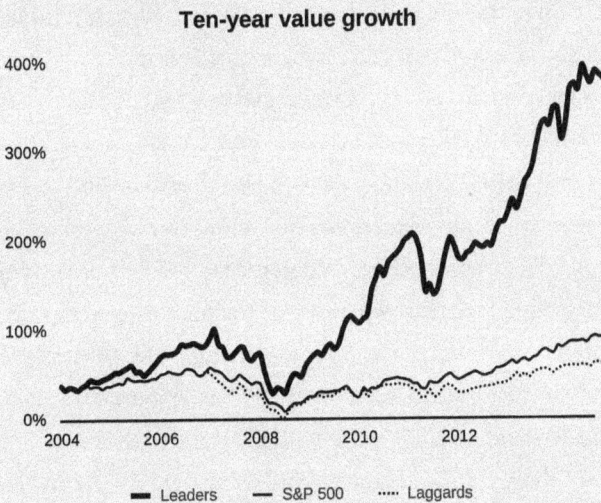

Figure 4: Firms leading in customer satisfaction beat their index peers by a wide margin in value growth[2]

A customer journey value stream is a flow that captures a specific customer persona's experience with that product or product portfolio over a period of time when they engage with that firm.

The customer-centric flow captures the exact interaction points that customer has with the firm. Generally, firms baseline that customer journey to use it as a tool to determine where they can enhance the customer experience. At the most basic level, the product leadership team will have key process flows documented.

Historically, making decisions about features to roll out or invest in was based on the firm's intuitive understanding of what the customer wanted—or simply their own preferences. While occasionally that strategy works, and the customer follows the business's lead, it is becoming more and more important in an enterprise context to get inside the head of the customer and make sure the organization is developing *what the customer wants* in a way that's congruent with *the way the customer is thinking about the job they are trying to do.*

The arc of that work begins with baselining the current journeys, then reimagining them to be either more operationally efficient, to be more feature rich, to address some gap, or to move the needle on some KPI of importance to the firm or customer. For example, take a case where a retail trading application's sign-up process demonstrates a heavy abandonment rate. The leadership team might choose to examine that customer journey to attempt to understand why, in particular steps, a lot of customers choose to exit the and instead contact the call center to finish. Is that a design problem where the steps are so convoluted the customer can't complete it digitally? Are the customers missing pieces of information that the firm could possibly help them capture electronically to bypass the decision to do it later, which leads to the customer failing to return to the process? This reimagination phase, which typically includes research and customer interviews, helps to inform the answers to questions such as these.

A real-life example of a customer journey developed by the Smithsonian Office of Visitor Services shows how one can look at their business capabilities from a purely customer point of view.

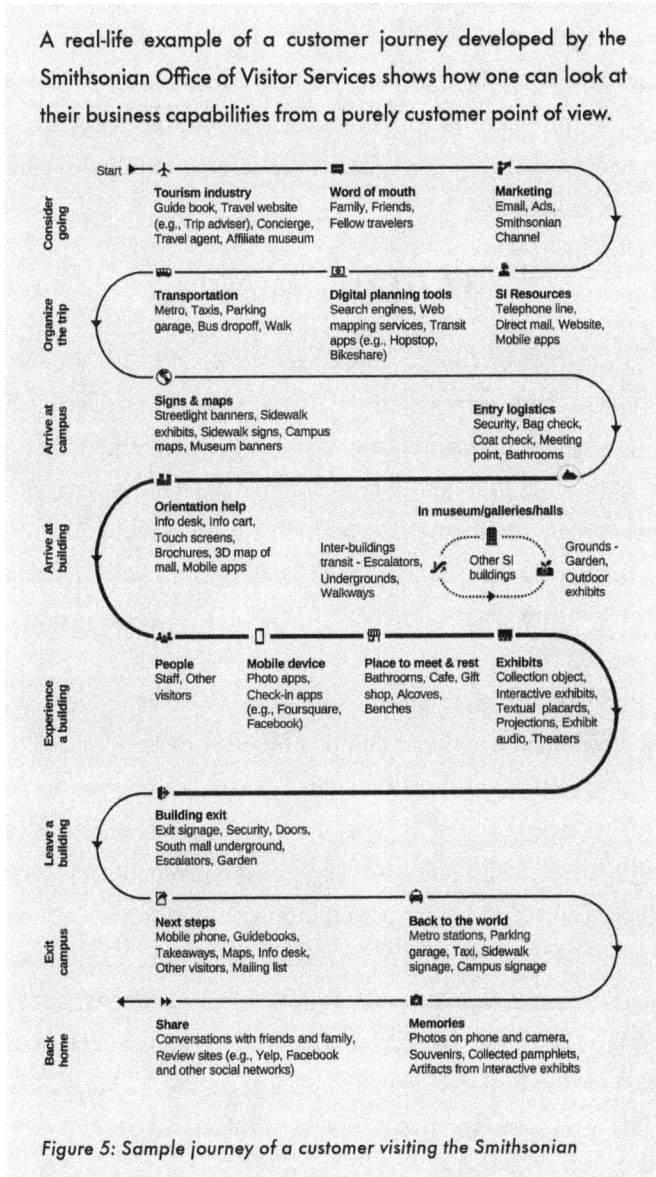

Consider going

Start ✈

Tourism industry
Guide book, Travel website
(e.g., Trip adviser), Concierge,
Travel agent, Affiliate museum

Word of mouth
Family, Friends,
Fellow travelers

Marketing
Email, Ads,
Smithsonian
Channel

Organize the trip

Transportation
Metro, Taxis, Parking
garage, Bus dropoff, Walk

Digital planning tools
Search engines, Web
mapping services, Transit
apps (e.g., Hopstop,
Bikeshare)

SI Resources
Telephone line,
Direct mail, Website,
Mobile apps

Arrive at campus

Signs & maps
Streetlight banners, Sidewalk
exhibits, Sidewalk signs, Campus
maps, Museum banners

Entry logistics
Security, Bag check,
Coat check, Meeting
point, Bathrooms

Arrive at building

Orientation help
Info desk, Info cart,
Touch screens,
Brochures, 3D map of
mall, Mobile apps

In museum/galleries/halls

Inter-buildings
transit - Escalators,
Undergrounds,
Walkways

Other SI
buildings

Grounds -
Garden,
Outdoor
exhibits

Experience a building

People
Staff, Other
visitors

Mobile device
Photo apps,
Check-in apps
(e.g., Foursquare,
Facebook)

Place to meet & rest
Bathrooms, Cafe, Gift
shop, Alcoves,
Benches

Exhibits
Collection object,
Interactive exhibits,
Textual placards,
Projections, Exhibit
audio, Theaters

Leave a building

Building exit
Exit signage, Security, Doors,
South mall underground,
Escalators, Garden

Exit campus

Next steps
Mobile phone, Guidebooks,
Takeaways, Maps, Info desk,
Other visitors, Mailing list

Back to the world
Metro stations, Parking
garage, Taxi, Sidewalk
signage, Campus signage

Back home

Share
Conversations with friends and family,
Review sites (e.g., Yelp, Facebook
and other social networks)

Memories
Photos on phone and camera,
Souvenirs, Collected pamphlets,
Artifacts from interactive exhibits

Figure 5: Sample journey of a customer visiting the Smithsonian

The base artifact of the customer journey is a customer-centric diagram that documents all the touchpoints where the customer interacts with the organization. For key customer jobs, understanding the process flow that customer goes through would be a basic level of documentation. Without this, a product team can end up deploying features or enhancements that don't directly serve a point of pain faced by the customer.

MATURITY MODEL

The principles we describe are hard pivots for many product leaders at large enterprises, and this mindset is best developed through a maturing process that is incremental and useful at each level, like any Agile methodology would deliver. For milestone purposes, we are prescribing three levels of maturity in developing this principle (and all others that follow), making it easier to imagine a continuum or evolution rather than often-tried and failed pivot.

There are many people who have asked us if they can improve upon one dimension of a business strategy, capability map, *or* customer centricity—not all three dimensions—while still achieving success in developing better products. It is our experience that this will result in an inferior outcome because each dimension has to be progressed proportionally so information uncovered about them can immediately be made useful for teams and businesses. As an example, if one developed a comprehensive business capability map and strategy dissemination but failed to improve the business architecture based on the learnings, they will not improve the product at a granular level.

	Level One	**Level Two**	**Level Three**
Business Strategy	Product engineering teams get frequent refresh of the business strategy from business and operations.	Product engineering teams can apply strategy and develop necessary product KPIs.	Product engineering teams can tie success metrics to key features and epics.
Capability Map	A comprehensive and documented view of business-defined capabilities exists.	A comprehensive breakdown and documented view of business capabilities into product and operational capabilities exists.	Product and operational capabilities are refined into a feature library that is actively used to inform roadmap and strategy.
Business Architecture	Product engineering teams have documented customer flows.	Product engineering teams can leverage research to further detail customer journeys.	Product engineering teams have developed value streams and have the capability to make optimizations and improve flow.

Level One

BUSINESS STRATEGY

At level one, a leader educates their team to increase their awareness of the company's business strategy. This high-level understanding enables the team to play an important role in the company's success. Regular meetings with the business leaders and customer operations teams should be conducted to gain this knowledge. Quarterly planning meetings offer a natural opportunity for regular meetings. Most modern scaled Agile frameworks include this topic as a key portion of any planning agenda. For example, in SAFE's Program Increment Planning (PI Planning) sessions, this topic is a day-one agenda for the entire product team.

CAPABILITY MAP

When it comes to business capabilities, over the years we have observed a direct correlation between the strength of the product management discipline and the usage of business capability maps by product teams and the business results they achieve. As Peter Drucker has said, "Efficiency is doing things right. Effectiveness is doing the right things."

A strong business capability map provides a shared vocabulary for the product organization; hence, this should be developed during the initial stages of maturity.

A business capabilities map does not have to be a static view, as we have defined earlier in this principle for introductory purposes, though that provides strong enough context to understand the current state. Successful product leadership teams develop layers of heat maps, which allow a particular problem statement to be viewed holistically for strategic planning.

In the "Large Insurance Firm" example, we have highlighted an insurance company for which claims processing has been determined to be an issue due to: (a) legacy platforms, (b) manual processes, and (c) product gaps such that the organization cannot keep up with competition. We have developed a business capability map to start depicting the organization and categorization of business capabilities as seen in this example during the initial stages of maturity to try to understand the potential pain points.

Large Insurance Firm

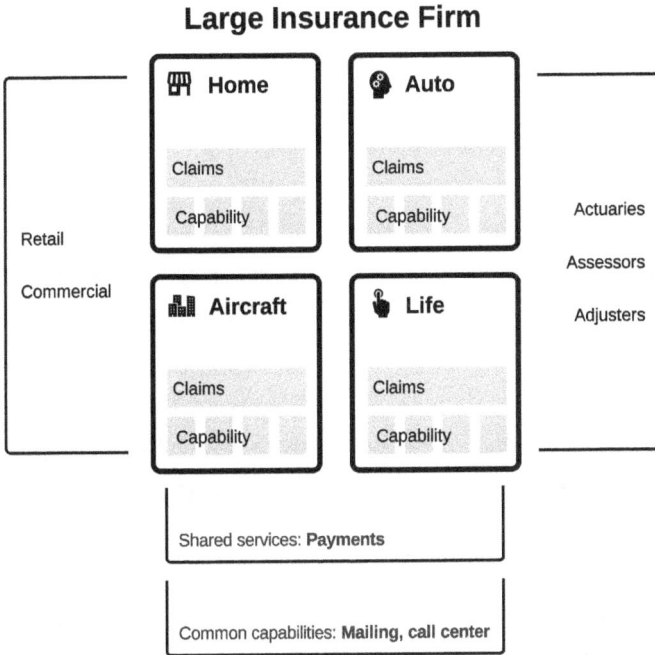

Figure 6: Sample insurance firm business capability map

For product leaders, driving organizations' roadmaps through an enterprise capabilities map or a set of business-specific capabilities maps allows decisions to be made with the full context and impact, and with economies of scale. This provides strategic program-level, board-level, and competitive views for meaningful conversations.

We are cognizant of the fact that people are busy and product leadership manages large teams. Fires pop up, demanding to be put out, and problems need to be solved quickly and efficiently. While it takes time and effort to create, a business capability map facilitates decision-making. When issues are confronted without full context, decisions made from that perspective will carry the burden of that lack of context, leading to misdirected hustle.

Even product leadership teams who presume to know their business context will benefit by first answering a few key questions to provide insight that may not have previously been considered.

The first question to ask is, **"Why do customers come to us as a business?"** The answers to this question will unlock the company's top-level capabilities, which can be included as simple boxes on a business capability map.

Next, they should ask, **"To perform those business functions, what capabilities do customers expect us to have?"** A customer might come to a well-known stock brokerage firm to trade a stock, but to do so, there also needs to be capabilities to open an account, move money, and receive statements. That customer isn't coming to this brokerage for great-looking statements, but they will expect that dependent capability to be practical and performant.

Finally, they ask, **"What do our target customers go to a competitor for, instead of coming to us?"** A complete capability map contains not only the functions the specific organization currently performs, but also the capabilities built into the industry and customers' expectations. For example, customers are not able to buy Bitcoin on E-Trade, so they may instead go to Robinhood because it allows for both trading and Bitcoin. Trading cryptocurrency should therefore be an industry function on E-Trade's capability map, but because they don't yet have that capability, it would be included in upcoming capabilities.

After answering each of these three questions, the product leadership team will have a high-level grouping of features that map out the business capabilities.

Whether they commission a team to run a project to complete this business capability map or a newer member of the product leadership team conducts conversations with various people as they step into their role, it's important that this capability map is a live document shared with their team so the knowledge can be shared. Ultimately, this document should be up on the wall in a visible place for the whole team to see, thus promoting its relevance and completeness.

In addition to creating the physical document, there are intangible benefits to performing the steps of this process of building business context. These include greater collaboration with partners, employees, and customers; better understanding of the company and its landscape; and increased insight into the people, technology, stakeholders, and data involved—all of which contribute additional context about the specific business as well as the larger industry.

After this initial data-gathering stage, level two takes these capabilities deeper to lay out the relevant sub-capabilities.

BUSINESS ARCHITECTURE

At this first level of business architecture, the product leadership team will determine whether the organization has documented journeys for their most important customer personas. If the answer is no, then the first step is to create those artifacts, at least at the most basic level.

The user experience or customer experience team should have an artifact that illustrates the customer journey so the product leadership team can use it to educate their team on the journey. We acknowledge, however, that not every organization is going to have this luxury. In that case, the leadership team can, at a minimum, document the process flows of how customers are using their products without advancing further.

Having simple, clear, key customer flows will help ground the team, keeping them on track and focused. These simple flows, which demonstrate how a customer reaches the final value offered by the firm, begin to plant the seeds of measurement and improvement that can take product teams to their next level of evolution.

Level Two

BUSINESS STRATEGY

As the teams mature, an evolution of the understanding of the firm's business strategy means developing a deeper under-standing of the KPIs that help the firm measure success against that strategy. A solid foundation of measurement leads to better business outcomes. Whether looking for revenue, profitability, customer churn, or other results, a strong understanding of the numbers can help guide decisions. A recent assessment by Bain & Company and Google found that of the 600 companies assessed, the 100 companies highest on the measurement maturity curve were four times more likely than those on the bottom to exceed business goals, grow revenue, and increase market share.[3]

CAPABILITY MAP

The business capability map evolves at each level by adding a layer of depth that demonstrates another element of capability. At level two, a hierarchical organization makes the business context information logically accessible and comprehensive without creating duplications.

For a wealth management firm, for example, it may be determined that customers come to an advisor for help with planning; thus, planning is a level-one capability. Underneath that main capability, a set of sub-capabilities can be found: investment planning, retirement planning, small business planning, and tax planning. Further, inside tax planning exist even deeper sub-capabilities, such as tax analysis, tax estimation, various tax strategies, and scenario modeling. Performing this exercise shows how to bifurcate a capability down to a fairly low level of detail.

A specific level-two capability map for insurance premium pricing is depicted in the example.

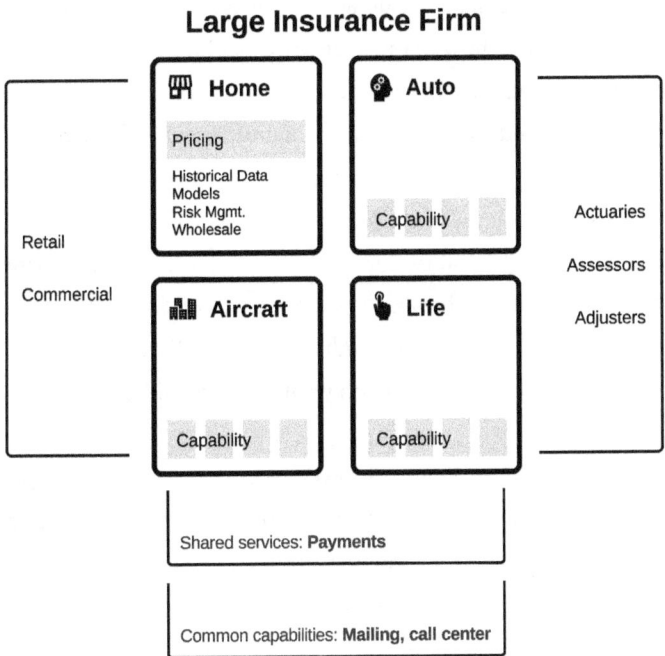

Large Insurance Firm

	Home	Auto	
	Pricing		
Retail	Historical Data		Actuaries
	Models		
	Risk Mgmt.	Capability	
	Wholesale		Assessors
Commercial			
	Aircraft	Life	Adjusters
	Capability	Capability	

Shared services: **Payments**

Common capabilities: **Mailing, call center**

Figure 7: Business Capability Map with detailed (Level II) pricing sub-capabilities

A deeper assessment of the capabilities in a certain area may be prompted by the kickoff of a key initiative, regulatory change, responding to market shifts, or addressing customer feedback. Additionally, as the product leadership team continues into discussions for other principles and possibly determines additional capabilities in their portfolio, they can add it at level one or flesh out an existing capability to reflect the appropriate depth.

For large organizations, it can be easy to become bogged down with the details of the capabilities map. Some large organizations would have hundreds of capabilities, each of which could be broken down into one to ten additional sub-capabilities. A regular struggle within this level is how far to drill down, as well as how to organize the capabilities and sub-capabilities to ensure harmony of the elements within capabilities. Each level needs some semblance of uniformity and balance so large items are not buried deep in the tree nor are minuscule items listed at level one. Different buckets can be used when categorizing information, and determining which is appropriate can lead to philosophical debates. The levels of classification are fairly straightforward, but product leaders will need to have a more in-depth conversation on what sub-capabilities should be classified as.

The level one categorization can seem easy when the major capability categories are large and most people agree that it is an accurate description for the product. But once the teams get deeper into sub-capability definition, there may be some disagreement with where specific abilities fit within this system or how they relate to other capabilities.

As an example, if we looked at a sub-capability like

"Historical pricing data for insurance quotes" (refer to diagram in Figure 7), an argument could be made as to whether that sub-capability lives within each line of business (Home, Auto, Aircraft, etc.), or whether it should be a shared capability. The implication of how the sub-capabilities are organized has a corresponding impact on people, systems, and processes that will support those capabilities. To get through these conflicts, consider the realities of the organization as a starting point. Over the long term, inefficiencies observed during this categorization exercise can serve as an impetus for subsequent organizational realignment of capabilities.

For some industries, standard industry models can be leveraged as a base starting point to develop a firm's business capability map. Some examples of industry capability models include ACORD framework for the insurance industry and BIAN for banking and financial services.

The best way to determine that a capability map has hit its value level is if, at a minimum, it has all the information pertinent to the organization at that point in time. If a product leadership team receives a directive from the CEO, saying, "We face increased scrutiny from regulators if we don't have appropriate privacy tools in the app," they will need to evaluate their landscape on privacy. When their team begins to deep dive into the sub-capabilities within the capability of privacy on the map, how can they determine when they have accomplished their goal? They can be sure they're in an optimal place when they can point to the appropriate sub-capability that speaks to the problem they're trying to solve. As they attempt to shake things loose in those meetings and discussions, the product leadership might ask, "What's something that everybody's doing from a privacy standpoint? What's something extra valuable that we could add?"

BUSINESS ARCHITECTURE

Customer journey and enterprise value streams evolve as well here in level two.

Once product leadership teams have simple, rudimentary process flows, they can advance in this area by creating comprehensive customer journeys and personas for key product journeys. Comprehensive journeys meet the industry definition standard of a customer journey with a set of attributes that journey documents, but additionally they focus on reimagining future opportunities on that journey. This level relies on the skill and ability of the team to understand the customer, know how to talk to the customer, observe the customer's behavior in using the product or service, and then be able to elicit value-driving opportunities—knowing how to calibrate those observations and that research gathering in order to capture that next evolution of opportunity.

This creates a rich set of data that backs up why, for a particular release, the product leadership team may decide to organize the priority of the functional delivery in a specific order. The customer experience team will be able to identify a feature someone might think would be frequently used, whereas in reality the data show that customers do not use that capability nearly as much as anticipated. This will assuage some of the opinion friction out of scoping decisions—it is not just someone's opinion; it is based in fact, supported by research.

Level Three

BUSINESS STRATEGY

At the most advanced level of maturity, business strategy becomes measurable to the teams. This measurement of strategy and associated value provided by the product teams percolates down from the higher level business investments all the way to product level epics and, in some instances, features as well. At this level, a team is able to make conscious epic and feature trade-off decisions based on the projected impact on the KPIs important to the firm.

Scrum Inc. suggests the following framework to help product engineering teams assess sources of business value:

Market Value

Will this feature allow us to:

- *Sell more units?*
- *Charge a higher price?*
- *Reduce the cost of providing the product/service?*

Risk Reduction

How will completing this story allow us to:

- *Develop or refine hypotheses about the market?*
- *Prove technical assumptions?*

Capability Building

Will completing this story:

- *Enable our team to do something we couldn't before?*
- *Reduce or eliminate the need for low-value activity?*

Effective feature prioritization can deliver radically more business value over time. At this advanced stage, a team is able to use epic net present values (NPV) and estimated story points to encapsulate its return on investment (ROI). This method encourages product engineering teams to think in business impact terms and, with reflection, improve their prediction accuracy over time.

CAPABILITY MAP

The capability map is a model of an organization's ability to successfully perform a business activity. It helps organizations understand the capabilities they can develop through executing their specific business model and makes for more seamless decision-making in executive circles.

In this advanced level, organizations will evolve individual sub-capabilities built in level two into more detailed feature maps, a rich set of rules, and feature descriptions that describe the product capabilities. Feature maps—deliverables typically associated with many Agile frameworks, such as Scaled Agile— are mind maps of features built into a product. Generally, these are developed in association with user experience, business specialists, and product engineering teams. Artifacts include specifications, wireframes, and high-level requirements, which will then be used in Program Increment and sprint planning.

Feature maps are crucial to institutional memory. Without these, organizations tend to lose the knowledge associated with product, ignore caring and feeding of features, and never iterate on product.

BUSINESS ARCHITECTURE

Finally, the last stage of the customer journey occurs when, instead of looking at the customers as one collective whole, the product leadership team is able to advance how they segment customers. Whereas previously they may have talked about the customer as a singular, now they can look at the different types of customers.

In a finance firm, for example, the three customer types might include: a young investor who loves to trade, reads the news, is in the market pretty frequently, and very savvy but doesn't have a lot of money; somebody in the middle who is more risk averse and who has a family, so they're more concerned with wealth building toward objectives like college or retirement; and finally a third segment of customers who are very close to retirement and are not at all interested in taking on any risk at that point, but who are more interested in drawdown strategies and advice on transitioning out of earning and saving to spending. The "customer" as a singular term now gets split into this segmentation, with unique journeys by each of those customer segmentation points. Based on the economic value of each of those customer segments, leadership may prioritize and sequence features differently.

One of the most important tools within business architecture[4] is value stream mapping, which captures an organization's processes for creating a product. This helps managers visualize the features, interface, or services that add value to customers and those that don't directly impact them at all. The commitment needed to capture this information can be significant: it takes staff time or external consultants' input (or both), but the payoff will help the many different parts of a company work together more efficiently and better serve clients.

Value stream mapping has become one of the primary ways

companies combat wastefulness; by visualizing their process for manufacturing products, they're able not only to identify where resources are being wasted, but also to prioritize areas that could use improvement. Because creating comprehensive value maps can take a significant effort, some organizations choose to begin with the core customer journey value streams before expanding those value streams to include operational and development value streams that affect the customer journey.

The origins of value stream mapping are often attributed to Toyota Motor Corporation. Value stream mapping (sometimes called VSM) is a lean manufacturing technique to analyze, design, and manage the flow of materials and information required to bring a product to a customer. The application of value stream mapping—also referred to as "visualizing" or "mapping" a process—isn't limited to the assembly line. Lean value stream mapping has gained momentum in knowledge work because it results in better team communication and more effective collaboration. Ultimately, value stream management helps to keep the focus away from individual opinions and prioritize actions based on the customer's perspectives.

Once the product leadership team has the list of pieces that constitute what their product can do—in a business capability map—they can add a value stream map across a process to determine how best to affect change. While the capability map is a static, unlinked set of features, different value streams can include various combinations of capabilities. Value streams are about more than simply writing down the process; successful product leadership teams will want to make them useful, linking them to the cost associated with each step and with certain operational metrics that allow improvements to be made and measured throughout the value stream.

Considering the value stream perspective is a critical step because it is rare for customers to view a capability in isolation. Those customers are attempting to complete a job, and that job requires them to touch many capabilities; thus, their satisfaction is going to be dependent on the experience of performing that job. Optimizing the value stream allows product leadership teams to ensure that those jobs are easier to do, faster, or less expensive, and therefore differentiated in the industry. For example, a client of ours wanted loan applications to be approved within two hours, with a human touching it—in most cases, a credit officer. This capability development required more than understanding the process flow. We had to re-engineer the value stream by performing automated underwriting, exception processing, and escalation management.

In another example, a customer coming to an advisor for tax planning touches that sub-capability but, more, they touch a value stream that spans everything from the time they ask for advice through until the moment the advisor is able to charge for a proper planning session. The firm attempting to optimize that value stream to reach a direct business result will want to evaluate the entire flow from setting up the appointment, to conducting the meeting, to transferring necessary information to advisors, to conducting tax planning internally, to vetting it and presenting it to the customer. That set of variables can all be tracked to measure the efficiency, the error rate, or any issues across those steps.

Value stream maps not only show the value to the customer, but also bring value to the business because they demonstrate what the company is spending and where. The product leadership team can then show their management that they are working on a program to save money strategically, as well as demonstrating the projected ROI. A value stream map without the capability map of levels one and two doesn't have the same business value because there are no processes to tie together. The business capability map facilitates the creation of the value stream; the product leadership team can superimpose the value stream over the business capability map to clearly see the series of connections.

At an advanced level, the team can expand value stream mapping past the customer journey and also begin to include other internal development value streams that support that customer journey. Examples of this could include creating value streams for development lifecycles of the sub-systems that support that specific customer journey value stream. It could also include data operations that are necessary to enable epics and features for that customer.

COMMON CHALLENGES IN IMPLEMENTING THIS PRINCIPLE

We have found that there are generally four reasons why product teams may not currently embody the spirit of this principle or may run into challenges in implementing it.

Product engineering teams' mindset is that product roadmap is someone else's job.

One common challenge when implementing strong product management discipline is the mindset of the teams themselves. People inside those teams may believe that the thought process around the product roadmap or business strategy sits with other teams. While this may be technically true, it is a mistake for teams to think, "I don't have to think about it; that's somebody else's job—just tell me which feature you want, and I'll code it and hand it back to you." It is important for leaders to continue to push their engineering teams to broaden their horizons and understand the business context in which they work.

Product leaders assume that educating product engineering teams through artifacts and documentation is a waste of time.

Secondly, the product leadership team may encounter the mistaken assumption that these artifacts are unnecessary. They may not be encouraged to spend the time to build a detailed capabilities map, value stream map, or customer journey map. It's not yet common for organizations to invest in enterprise customer experience teams that have a focus on building end-to-end customer journeys and educating the rest of the organization on how customers are thinking about the jobs they want to do and where the pain points exist. The business architecture discipline is a multifaceted way to view the company. It's important for an organization to focus on investing their resources here in order to succeed long-term.

Organizational silos define what a product team can and cannot do.

The third challenge the leadership team may encounter is politics. People on engineering teams may feel that if they cross into asking questions about product strategy or about how the team came to a certain conclusion about a feature, it could be perceived as wading into somebody else's "turf." Employees may think that they just need to do whatever somebody else has already prioritized, without asking for deeper understanding of where their role fits in. This silo does not exist in smaller and more digital-native firms that enterprises are competing with; there is typically a symbiotic relationship between the product management team and the engineering team in terms of discussing creative ways to engineer the product. It is imperative for enterprise leaders to continue building collaboration across departments and functions. Fostering creativity, accountability, and collaboration will not only help the company grow, but allow employees opportunities to expand their horizons.

There exists a misunderstanding on how to be an effective strategy communicator.

Finally, there may also exist a misunderstanding of what it means to be Agile. Exemplified by the common phrase "pace over perfection," people may think being Agile means simply moving faster, without investing time documenting and creating these artifacts. They think they can be effective without all the foundational elements in place. It is a mistake to skip these critical foundational steps in favor of what seems like speed. In the long run, this could lead to hustle that has no direction and will not yield the desired results.

A Practical Example

The leadership team at a portfolio company of a private equity firm we worked with exemplifies a team that did not espouse a product management mindset from the top down. The team's lead architect pushed specific technology choices without consideration of the economic impact to the firm or the overall business strategy. The product management function was very disconnected from the main team that was building this new platform. The team also did not have documented customer journey maps and artifacts to serve as a sounding board for product feature decisions.

Ultimately, the program has been delayed by multiple years, significantly pushing back desired business results and any economic gain they may have hoped to achieve. Team members were unfortunately also let go to restructure the program against the new delivery time horizon.

When we were consulted, we had to shift the engineering leader's mindset to building something more useful and usable; we had to approach changes from a customer-centric point of view; and we had to ensure that the product being built made sense for this particular operations team. Because there were no artifacts and no clear connection to the business strategy, we also had to reverse-engineer these priorities. Although one sleeve of the multiyear program has gone live, the earlier dysfunction continues to have a lasting impact on this organization's ability to deliver the desired results.

CASE STUDY: WEALTH MANAGEMENT PLATFORM

A new head of product for an institutional wealth management firm inherited a large team, with twenty-six work streams, but no existing set of capabilities mapped out. The business identified several fundamental issues: collisions across teams on who owns what, challenges on the finance side in terms of where to make investments on features and capabilities, and not having a way to go to the customer and communicate around upgrades.

The head of product had an assortment of PowerPoints for various conversations but no central place to assign and categorize the feedback his team was receiving because they weren't being managed against a master capability map. The next time he stood in front of advisors for a quarterly update, he had to field complaints such as, "Half the features we discussed last quarter didn't make it in!" Or, "I noticed you added new features in this capability area, but why aren't we doing more in these other capabilities?"

These comments emphasized his realization that there were holes in the scope—and that he couldn't keep the pace of updates to the capabilities in this manner. That triggered him to create a detailed capability map for this product. After doing so, everybody had a copy of the business capability map; it went up on every wall and was posted in the team war room. The new product capability map has made it easier for the business departments to collaborate. They can now anchor their feedback more systematically and have a broader discussion about features that should be prioritized or invested in first.

His team was then able to use the business context diagram to move to Principle 2, which is system architecture, and map their systems by that capability. From there, the data gathering became significantly easier because there was a shared vocabulary across all the teams. In Principle 4, they were able to organize the stories and epics that the delivery teams were working on, all mapped up to the same shared capability maps. It became easier to track progress against key features that were communicated by the customer by capability.

The business capability map served as a north star for that business line in terms of their total lay of the land—like the general surveying the battlefield, they were able to gain a clear perspective of where they chose to invest and not invest, and the differentiators they chose to emphasize. With everything in context, this process led to a more harmonious organization of work across the teams and an understanding of how the capabilities fit together.

The more time spent on the customer journey and business capability map, and the more thorough these artifacts become, the more it helps facilitate deeper conversations. The business capability map is also the lead input for Principle 2, which asks how well all the company's systems and engineering infrastructure has supported or fulfilled these business capabilities. In Principle 2, we create new blueprints based on the business architecture determined through this principle.

KEY TAKEAWAYS

1. Every engineer needs to know the why, what, and how of the organization's business strategy and be able to apply it to everyday work product.

2. Technology to Product mindset change is not a pivot; it is an evolution, but there is measurable benefit at every stage.

3. Business Strategy, Customer Journey, Value Stream Maps, and Business Capability Maps are critical tools that help enable an engineering organization to develop a product mindset.

PRINCIPLE 2

Aggressively Attack the Weakest Link in the Product Architecture with Transformative Technologies.

"Like Janus, a software architect needs to be a keeper of doors and passageways, spanning the old and the new, incorporating creativity with sound engineering to fulfill today's requirements while planning to meet tomorrow's expectations."

—RICHARD MONSON-HAEFEL,
97 Things Every Software Architect Should Know

Evaluating product architecture with an eye toward attacking the weakest link demonstrates the ability to take a competitive and innovative product to the market. Although there is no formal definition of weakest links, major enterprises often use the term "legacy software," usually represented by software that is unable to support future business requirements, is not amenable to change, or does not have skills in the market to maintain effectively.

An IEEE study[5] showcases that seemingly simple and straightforward requirements often become unable to be replaced, not because of complexity of the requirement but because the systems that support these requirements receded into, using IEEE's term, the "operational shadows" of the organization.

To give an example of such "straightforward" requirements, the head of product for a brokerage company wanted to offer customers a deal to entice more high-volume traders to their firm: heavy traders would receive a discount on their monthly trading fees.

After examining their systems, however, it was determined that the technology could not support the capabilities necessary to implement this change. It was not possible to process the discount and configure the statements correctly because the legacy systems, as we have previously mentioned, are in the operational shadows. The technology owners and architects do not know how to tweak it, nor what impacts this change could cause to the rest of the organizational processes. This function obviously could not be performed manually, given the large number of customer statements delivered each month by the legacy system. It didn't look as though there was any way this company could efficiently offer this product feature to their most profit-making clients.

How could a team that had such a prolific product, used in the marketplace so successfully, end up in this situation where extensive investment is now needed? How did this system go into the operational shadows? This scenario happens more times than one might expect; according to the same IEEE study, only 25 percent of IT budget is spent on modernizing existing software.

THE IMPORTANCE OF AN AGGRESSIVE ATTACK

The clients in the above example would likely not have found themselves in that situation had they consistently evaluated their product architecture and aggressively attacked the weakest link, in this case the client billing, fees, and product configuration modules.

A landmark study by the US Department of Defense, quoted by IEEE, called "The Software Is Never Done," describes the process of constantly fighting to create legacy.[6] This principle outlines a system approach to not only avoid legacy, but to aggressively modernize and stay ahead of it.

Enterprise architecture is the blueprint of how an organization's business capabilities are supported by the systems of the product portfolio. Principle 1 created an understanding of the business strategy, of the capabilities provided for customers, and of the journey customers take when using the products in the portfolios supported by product leadership teams. Principle 2 now seeks to answer, "How do successful product leadership teams deliver those capabilities to those customers in line with that business strategy?" Ultimately, that answer lies in developing the blueprint of systems, subsystems, and modules a company has in place to support their business capabilities. For example, if the business capability map demonstrates that one needed capability is to support credit card payments, here, on the enterprise architecture side, the product leadership team will examine how credit card payments can be supported—through a partnership with MasterCard or Shopify, for instance.

Similar to the business capabilities map created in the previous principle, the product leadership team will now create a system capabilities map, which adds the layer of understanding which capabilities are being supported and *how* they are being supported. Some capabilities may have custom systems, while partners support others, and still others have teams manually grinding through them.

As a result of that mapping exercise, product leadership teams can begin to take a view of the capabilities that are being supported by the architecture and assess where the risks of technologies becoming obsolete occur or where opportunities exist for new features to invest in and build, thereby triggering a change for the organization. This may reveal that there are no systems currently in place to support desired capability X or that a cluster of systems are utilizing outdated technology, leaving important capabilities at a tech-debt risk (as seen in the example involving configurability of fees on a mainframe discussed at the beginning of this principle).

Once product leadership understands the system capabilities map (which we will delve into further in subsequent sections of this principle), the next step is to follow the principle of aggressively attacking the weakest link in the product architecture—by which we mean the most cumbersome or least flexible portion of the product feature set—with transformative technologies. Weakest links such as these are most often created through the inability to sunset legacy platforms that are core to the business, a risk avoidance culture of engineering organizations, and improper attention to product flexibility by product leadership.

We have found that few teams aggressively attack the weakest link in their architecture. Instead, that technology stagnates and festers while the product leadership team focuses on other more urgent functional priorities. By the time managing the weakest link becomes unavoidable, that architecture has become a formidable part of the organization's ecosystem, thus requiring a wholesale level of investment. As parts of the architecture become more and more calcified, it becomes more and more difficult to introduce new features without significant cost to implement. At that point, everything is slowed—completely counterproductive to the organization's goal of increased momentum.

Additionally, talent is scarce and developers may become dissatisfied working on a product if making even a small feature change is a chore and risky—even the ones that aren't particularly valuable to the business or likely to lead to significant customer delight—meaning that they have to overcome fear of making a change, make duplicative or debt-ridden changes, and perform extensive testing just to be able to roll it out. In this situation, the effort-to-delight ratio is skewed, leading to a loss of morale and attrition—with the eventual possibility of losing all team members who have a working knowledge of that older technology. On the other hand, introducing new technology and innovation to the product helps teams to feel fresher because they are constantly learning and they are able to see the positive impact on the customer and for the business in terms of new capabilities delivered.

The gap between the experience the customer wants and the experience the product provides will widen, and, over time, customers will go where that experience is faster or more intuitive. It is important to aggressively attack the underlying technology that might influence this experience—continually upgrading, evolving, and strengthening the architecture—in order to minimize this gap and, ideally, exceed the customer expectations.

It is impossible to attack the weakest link in the product architecture without a clear understanding of which link is weakest. Conducting an inspection is necessary to identify risks but also to establish the continuous nature of this evaluation. The more proactive and strategic a momentum leader is in establishing a continual process of evaluating the architecture of their product portfolio, the better positioned that portfolio will be to take advantage of new technology. If inspection becomes a quarterly activity, rather than one that is performed on an ongoing basis, it is easy for the evaluation to be pushed out further and further, which leads to the bind the teams are currently in, where they are forced to be reactive instead of proactive. Allowing architectural and technical legacy, or "debt," to grow over time exponentially compounds the difficulty in moving away from that data and reducing the reliance on those outdated technologies, thus inhibiting the team's ability to deliver innovation in the future.

In large organizations, transforming the product architecture requires multiple levels of vetting, compliance, and alignment—and it takes time. Most enterprises have enterprise architecture teams that govern the architecture to provide some level of uniformity and compliance across teams—and those architecture teams have their own orbits in which they run, with multiple product teams working across multiple portfolios. They are not able to predict the needs of every product across the enterprise. The more proactive a product leader can be in engaging those teams, in evaluating and keeping their products on the architecture team's radar, the more responsive and prepared the architecture team is likely to be, leading to more engagement and faster turnaround from those central teams in helping to facilitate the change.

Once the product leadership team has identified the weakest piece of their technology architecture, teams often become paralyzed trying to choose between replacement technologies that are either too cutting edge—leading to opposition to technology that is too new, too much risk, or too unknown—or overly conservative, exhibiting a bias toward proven technologies—platforms or frameworks with which they are already familiar so as not to have to hire or train people to support the new technology. When teams run into this point of drag, it ends up causing a loss of momentum. Thus, when making architecture technology decisions to support business capabilities, product teams must make these decisions based on an appropriate ratio of: (1) the ability of the organization to effectively support the technology; and (2) the expected future lifespan of that technology—with a bias toward pushing the envelope and moving the product team into the future.

COMMON CHALLENGES IN IMPLEMENTING THIS PRINCIPLE

Recurring process to identify architectural weak links are not established.

There is not likely to be an established, recurring process to determine the weakest link, as defined by this principle, or to evaluate where problems may occur. The traditional processes that exist are likely more reactive in nature, which does not provide enough of a runway to make transformative decisions.

There is unclear ownership of impacts to operational processes.

In addition to the technology changes the organization may need, business process changes—whether regulatory, operational, or support—can also be impacted by newer technologies. This can also lead to paralysis as teams believe modifying operational processes, for example, is not in their jurisdiction, so they are unable to move forward with the adoption of new technology until those changes are made by other groups or stakeholders.

As an example, one of our clients had customers who wanted to be able to digitally sign documents rather than having to print them out, sign the physical papers, and then fax or mail the documents. E-signatures as a product feature is widely becoming a common use; most countries and states accept e-signatures as official form of documentation. Product leaders who plan for such changes ahead of time and obtain enterprise-level approval are ready to take on such challenges.

However, a seemingly simple change, such as e-signatures, has multiple challenges in a complex product environment. Where will these documents be stored? Who will manage the regulatory requirements for information management (record keeping, destruction, and compliance)? Will e-signatures be accepted in all the countries we operate in? What happens when we do need a wet signature? What about cross-border transactions, which regime we will follow? How does Power of Attorney (PoAs) work with e-signatures? These are use cases and patterns that need to be solved prior to adoption.

In this example, with a customer-centric point of view, our clients proceeded with the changes in the jurisdictions where it was possible to obtain e-signatures exclusively as a pilot. Generally, how would product leadership stay ahead of these business and architecture requirements? How would enterprises support these? How long will it take to obtain approvals to use such tools?

Central architecture teams lean too conservative in an effort to mitigate risk.

We've often seen leaders struggle because they tend to defer to the centralized enterprise architecture teams, which in turn tend to lean conservatively toward leveraging already approved patterns and technologies; centralized teams are trying to manage risk. In large organizations, enterprise architecture teams hold the keys to the kingdom in terms of making foundational software, platforms, and tools available to product engineering teams. Given the operational and reputational risk associated with large organizations, enterprise architects follow methodologies such as TOGAF and ITIL, to enrich the technology capabilities and architecture library of the organization. Product engineering teams generally misconstrue the strict adherence to these methodologies as red tape[7] and instead want to focus on the latest and greatest technologies without allowing enterprise architects to vet the suitability from all dimensions of a chief information officer's (CIO) and CTO's organizational responsibilities (such as vulnerability assessment, scalability, legal liabilities, etc.). Product leaders may be risk averse themselves, not wanting to take accountability for any downside to pushing the envelope to new technologies and patterns that aren't blessed yet. Conversely, successful product leadership teams take more of an ownership—or at least a co-ownership—role (which will be discussed in greater detail later in the principle) with enterprise architects and are able to demonstrate leadership in taking the product where they want it to go.

Ownership of driving change is diffused when the underlying technical modules are used by multiple product teams.

Additionally, the ownership of driving change may become diffused if the pattern is used by multiple products. To return to the example with electronic signatures, it is likely that multiple products within that financial services firm would benefit from providing the ability to sign digitally. The owner of one product within that organization may be ready to move that product down the roadmap with digital signatures, but at the organizational level it may be unclear who owns signatures—it is unlikely that one product will move forward with digital signatures while the rest are still on paper. If there is no single owner for this new feature, it may also be unclear how to facilitate where to begin making the change. As will be discussed in an upcoming section of this principle, the push here will come from the business case and the business strategy.

Starting product implementation activity without foundational architecture in place.

Finally, we've also encountered product teams that have preemptively started the implementation activity without having these foundational architectural decisions in place. In these cases, momentum is certainly impacted downrange, especially if changes are not mapped to Principle 1, i.e., the business context, to ensure they are connected to some business benefit or product capability benefit.

Those who have not fulfilled the business capability map from Principle 1 will find that, even if this systems capability map is created, it will not provide direct and impactful value to the leadership without business context because it doesn't show specific capabilities the systems were supporting and how they are strategic to the business strategy. The product leadership team may refer to platforms and applications they have in the portfolio—but *why* do they have them? What is buried in those systems? What business purpose are they serving? Without mapping to business capabilities, the product leaders don't have context to be able to make strategic decisions for their products and obtain funding and leadership buy-in.

HOW TO AGGRESSIVELY ATTACK THE WEAKEST LINK

In large enterprises, architectural standards are passed down from the top. Generally, enterprise architecture teams leveraging methodologies such as TOGAF, Zachman, or ITIL will drive architectural capabilities of the organization. In addition, reliance on centralized platform and software usage, either commercial or open source, drives architectural pattern realization for production teams. How can a product engineering lead get ahead or stay in tune with these teams?

> A tool that is generally prescribed by these methodologies is called "solutions repository." Artifacts within solutions repositories are referred to by many names. For example, within TOGAF, the deliverables are SBBs (solution building blocks), and other methodologies refer to them as enterprise perspectives.8

Our principle of aggressively attacking the weakest link is to drive solution building blocks; an example could be, if a product needs to process large volumes of data in real time, in the near future, product leadership teams need to work with the enterprise architecture and platform teams for necessary platform selection and approvals.

Solutions repositories are actionable engineering perspectives on how product technology capabilities should be evolved over time, with tools, frameworks, patterns, and practices to enable future capabilities and to retire legacy.

There are three parts to our plan for how product leadership teams can aggressively attack the weakest link in the product architecture.

Part one of this process is for product leadership teams to establish the cadence of holding Forward-Looking Architecture

Risk meetings with their teams, the enterprise architecture team, and potentially additional people who can provide the business perspective.

The second part is to have the right mindset going into that meeting. Leaders should co-own the product architecture with the enterprise architecture team, making sure that team is invited to this meeting and involved in planning as equal partners so that they are advocates of the desired changes, enrolled in this new strategy, and can help guide the architectural direction for the product.

Finally, part three consists of managing the "cone of architectural uncertainty," which comprises three Zones.

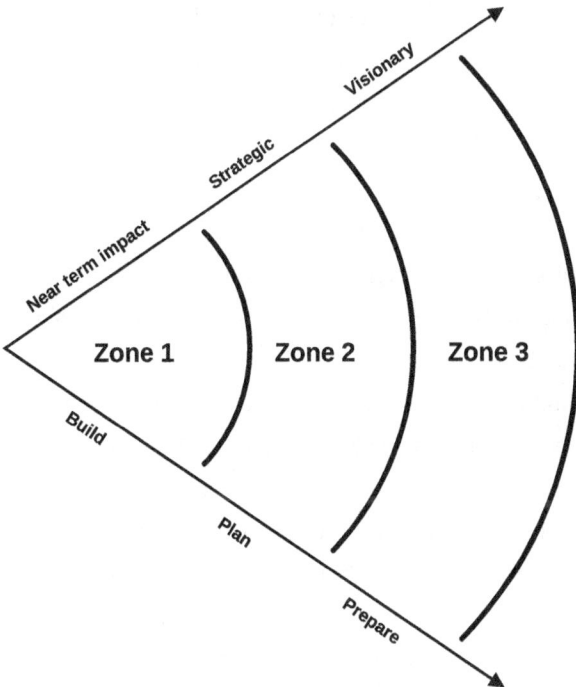

Figure 8: Cone of architectural uncertainty

- **Zone 1:** *Capabilities that have already been built and are live but facing challenges, such as a component that fails every time there is a spike in trading volume or certain capabilities that customers complain are difficult to use or very error prone. This could include subcomponents that are hard to maintain.*

- **Zone 2:** *Capabilities that are known and in the near-term plan but not yet built. The team has identified multiple solution options that need to be closed on in an enterprise architecture plan (EAP).*

- **Zone 3:** *Capabilities that are further out in the product-time horizon and need far more advanced technologies to support.*

The first several iterations of architecture capabilities discussions will most likely focus on Zones 1 and 2. As those are managed, the focus will shift to a rhythm where the team can spend more time discussing Zone 3. (Alternately, one meeting of the quarter could be designated solely to look at these way-out capabilities; we will discuss more of the schedules in our kaizen principle later in the book.)

A maturity model evolves as a result of these meetings. At a basic level, the product leadership team will create a systems context diagram, which is then mapped to the business capabilities map from Principle 1 (this is reflected in the maturity model of systems context in the next section). At a minimum, the team should identify gaps and pain points and may begin to consider solutions, but they still predominantly rely on pre-approved patterns and technology. The enterprise architecture team is involved, but the product team is not by any means the "tip of the spear" in terms of adopting new technology at the organization level.

The next level of evolution occurs when the product

leadership team's conversations with the enterprise architecture team become more proactive. Rather than asking what technologies are approved, leaders are thinking further ahead about multiple solutions (in Zone 2) so they can involve the enterprise architecture team on decisions to be made down the road. They are in the mental state of biasing toward newer technology and helping to influence their own roadmap. At this stage, the product leadership team may also be contributors to new patterns or new products that, with their help, have now been adopted at the enterprise level. An enterprise architecture plan is also established.

The final level of maturity is reached when the product leadership team has enough control over the architecture, and they are proactive enough that they consistently meet to discuss and plan for Zone 3 horizon capabilities and feel confident the product is positioned to be able to deliver those capabilities. Because the focus of Zone 3 is so far out, it's unlikely that even the enterprise architecture team will have proofs of concept or opinions on what technology to choose, but the product leadership team has established a degree of trust such that they are consistently able to choose from those more far-range technologies.

MATURITY MODEL

The principles we describe are hard pivots for many product leaders at large enterprises, and this mindset is best developed through a maturing process that is incremental and useful at each level, like any Agile methodology would deliver. For milestone purposes, we are prescribing three levels of maturity in developing this principle (and all others that follow), making it easier to imagine a continuum or evolution rather than often-tried and failed pivot.

	Level One	Level Two	Level Three
System Capabilities Map	A comprehensive and documented view of systems exists, and technical capabilities are mapped to business-defined capabilities.	A consolidated view of the metadata, pain points, and architectural attributes of system and technical capabilities is documented.	A formal Architecture Assessment and Planning process is in place to aggressively transform weaker components of the system capabilities map, as defined by business strategy and value stream assessment.
Solutions Repository	Established operating model to work with enterprise architecture team to resolve tactical product gaps and operational pain points.	Established change management process to choose and implement solutions for strategic business initiatives.	An architecture change management process has been established to select technical capabilities to enable visionary solutions for the industry.

Level One

SYSTEM CAPABILITIES MAP

In the first level of maturity, it is prudent to create an inventory of software applications from a portfolio point of view. What are all the platforms and systems in use across the board?

This bird's-eye view of the systems and applications and how they are supported can be a simple Excel spreadsheet listing useful metadata, such as the name of the system, who owns it, how long it has been in use for the company, and what kinds of technologies are used to build it. Every system in the portfolio has products associated with it, people who support it, and potential points of failure, so this gives some color in terms of whether those systems are in line with industry standards. At a glance, it can be determined whether there are too many systems, too few, or whether (as an example we've actually seen) half of the systems were built prior to 1950.

We've seen some clients struggle at this level in gathering the inventory itself. It's rare for organizations to be at even this preliminary level, but if they have a basic inventory, it's often out of date. A time investment is required to determine all the systems in a portfolio of products, but it isn't difficult. Someone in the data center can provide a list of applications and what they do. From there, the product leadership team can also add information from their management information systems (MIS). An organization's enterprise architecture team or architecture review board should be able to provide this information, at the very least at this basic level of detail.

From there, this simple, raw inventory can be separated by usage and mapped to the business capability map created in Principle 1. If the product leadership team has already been through the process of iterating through the first and second level of business capabilities in Principle 1, it will be significantly easier to accurately place these systems now.

Suppose an organization identifies billing as a necessary capability. In the initial step of this inventory, the entirety of the technology landscape of billing features used by that product or product portfolio team will need to be determined and documented. Here, specific applications will be linked to the appropriate capabilities—allowing the product leadership team of that organization to see, to continue the hypothetical example, that they have four different internal platforms for billing and instigating an investigation into why they have those four systems, if they are all necessary or if they are duplicative, and how maintaining those four systems impacts the organization.

If a product leader delegates this completely to their technology counterpart, without linking systems to business capabilities, the conversation will always start with two distinct and disjointed topics. When they are mapped together, however, so that enterprise architecture is presented hip to hip with business context, it allows all relevant parties to be of one mind.

Product architects often make the mistake of skipping this step and fail to work alongside enterprise architecture teams. They create the inventory and then operate independently, without the relevancy given by business context. This link is important, however, because it allows the product leadership team to focus on areas of concern or the potential impact of making a change. If, for example, their business context heat map shows issues in billing or client onboarding, this level shows the technology linked to those areas so it can be determined if anything in their systems and application portfolio is potentially the cause of those issues. Similarly, if they need to retire a product from their portfolio, the link between business context and enterprise architecture will show the impact that move would have.

This level tends to be the biggest pain point of Principle 2 as it is where the product leadership team may experience some pushback from teams who are skeptical of the need for this inventory.

With the understanding that successful product leadership teams have the managerial capability to overcome or avoid the resistance issue, the genuine obstacle here is that people are needed to conduct an inventory—and people are busy. Yet the context can *only* be provided by the people in the organization. A tool can determine that there are ten systems in place, but only a human being can explain that four of those systems don't do anything and the other six are the main ones used. It falls to the product leadership team to allocate and assign the time as well as provide the vision to the team about *why* they are being asked to do this and what purpose it is going to serve. Without providing the vision, people are simply being asked to do work with no end in sight.

Product leadership teams will run into a gamut of team members. Some will be volunteering with the necessary information, and others may view that information as their own intellectual property, which they're reluctant to share because they don't want to hand over the keys to the kingdom. When walking into working sessions with some of these senior roles and dealing with a variety of portfolio of teams, including people who have attached themselves to a product and whose identity may be associated with that product, there's a calibration to be mindful of where those types of pockets exist. Without that sensitivity, whatever team is conducting this exercise will not be able to get the complete set of information because they're going to run into team members putting up barriers. This is where bringing in an outside party can provide an advantage in inventorying applications. People are psychologically more receptive when it's framed as taking a task off their plate because their focus is needed on their primary undertaking.

Once the catalog has been mapped to product capabilities, the product leadership team can begin to determine certain areas of improvement or new product initiatives, taking a particular capability or sub-capability to the next level of this maturity model.

SOLUTIONS REPOSITORY

While system capabilities are being mapped and documented, it is important to make progress on setting up a governance model to attack the weakest links of the product architecture. There will be components that will fall into Zone 1 of the product's weakest-link cone of architecture we discussed, and leadership teams need plans to resolve these immediate issues.

In order for product leadership teams to make measured progress toward their own goals, it is critical to understand the architectural governance model of the organization. Leaders must own their modernization future; it won't be solved by someone else. In some enterprises, chief innovation officers or innovation Centers of Excellence (CoE) will make this process easier for product teams to try out new patterns. During level-one maturity, it is important to find the right stakeholders and partners to start moving away from legacy efficiently and in a timely manner, in alignment with enterprise architecture.

Level Two

SYSTEM CAPABILITIES MAP

In level two, issues determined from the business context can be examined in terms of enterprise architecture to create a heat map of specific capabilities and the systems that support them, leveraging the capabilities map created during level one.

Here, more detailed questions will be asked in order to identify the next set of artifacts. These may include:

- *What kind of application is it?*
- *Did the organization buy it, build it, or outsource it?*
- *How was it built?*
- *How long ago was it built?*
- *What kind of pain points or challenges is the organization experiencing?*
- *Why are these issues occurring?*
- *What decisions need to be made, given those pain points?*
- *How big is the organization team that supports this capability?*
- *How would it work to support a new product, if necessary?*

The answers to these questions will lead to the individual application architecture as a deeper part of the overall enterprise architecture. At level two, the product leadership team can answer critical questions such as these that they wouldn't be able to answer at level one. It takes more time to go to this level of depth, however, so prudent teams will want to choose their areas of interest rather than going deep for every area previously mapped.

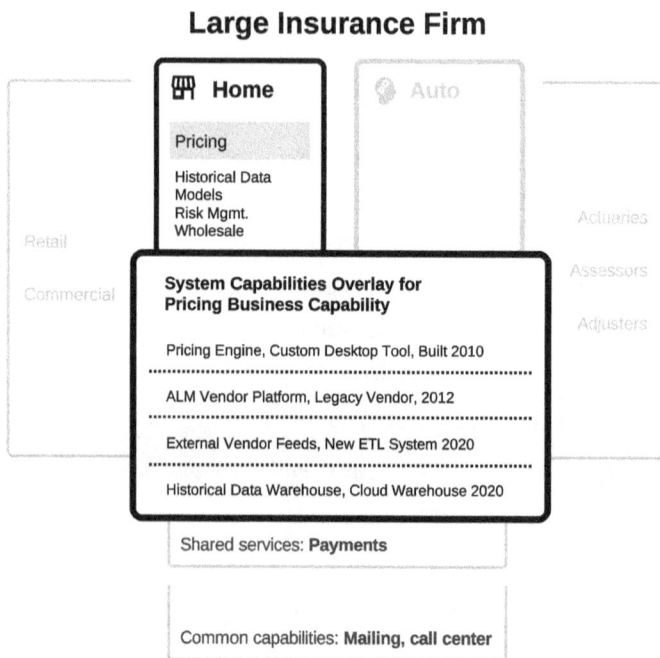

Large Insurance Firm

Home

Pricing

Historical Data
Models
Risk Mgmt.
Wholesale

System Capabilities Overlay for Pricing Business Capability

Pricing Engine, Custom Desktop Tool, Built 2010

ALM Vendor Platform, Legacy Vendor, 2012

External Vendor Feeds, New ETL System 2020

Historical Data Warehouse, Cloud Warehouse 2020

Shared services: **Payments**

Common capabilities: **Mailing, call center**

Figure 9: Systems capabilities overlay for pricing business sub-capability

To tie this concept back to a previous example, assume that the chief operating officer (COO) asks, "Why do we have four claims platforms? And what would it take for us to reduce costs by integrating into one?" It is necessary to go deeper for the claims capability and these four systems. Perhaps it is determined that there are four claims platforms to address property, casualty, vehicle, and umbrella insurance. Not all claims are the same, so each platform has different functionalities, and the story stops there. If, on the other hand, it's determined that the organization has bought six other companies in the past fifteen years, which has led to having multiple platforms performing similar functions, that would be a place for the product leadership team to say, "You're right, that doesn't make sense; let's consolidate."

The product leaders may go into level two for a current pain point and, later, when another pain point arises, they are likely to find it necessary to go to this level for a different capability or sub-capability. They may also have certain mandates from their CEO, COO, or business customers that they're trying to target. Those are reactionary, going deep on those particular areas, but there are proactive milestones as well, where some of the expected downtime certain teams have can be used to advance the ball in all of these principles. We'll examine this concept in more detail during the continuous improvement principle section.

SOLUTIONS REPOSITORY

In this level of maturity, attacking the weakest link is no longer considered tactical issue management, but instead becomes more about aligning enterprise architecture, product, business, and engineering teams to the business strategy.

In many cases, this level is about planning and budgeting for sunsetting legacy platforms. Product leadership teams will have the highest return on investment and better customer experience with these programs, which are also the most challenging from a people and culture point of view. People challenges occur because legacy systems are tied to jobs and these programs are considered politically challenging. Cultural challenges occur because these transformations require breaking silos and working together in a product-oriented mindset. McKinsey calls it upgrading the hardwiring of the organization.[9] From the architecture point of view, in order to alleviate these burdens, it is critical to align on a governance model discussed during level one.

Building upon level one, with an actionable governance model aligned to the enterprise, level two becomes about identifying solution patterns for the Issues and Strategic Zones (part of our cone of architectural uncertainty) and start replacing solutions with confidence that the enterprise will support the initiatives. We see product leaders end up spending large amounts of time during the "transformation or digital product" build period on these activities, and it is often too little, too late. It cannot be expected that such fundamental decisions will be available at the same time of program execution.

Level Three

SYSTEM CAPABILITIES MAP

A product engineering team should strive for level three when it is determined that the product leadership team should actively transform a portion of their business capability, not just replace the foundational architecture. This level creates a point of view on moving certain systems in a specific direction, rationalizes decisions by creating more detailed documentation on why this particular capability is not efficiently managed within the organization, and develops actionable solutions.

Similar to Principle 1, organizations will have the necessary foundation at this level of maturity to improve customer experiences, add value through reengineering from value stream assessments, and really bring innovative and differentiating solutions to their customers.

SOLUTIONS REPOSITORY

In this level, the focus is on adding new business capabilities and exploring value propositions that are truly industry first and visionary. Large organizations tend to skip to Zone 3 (Visionary initiatives) without successfully implementing Zone 2 (Strategic initiatives), which will require engineering teams to manage a large amount of legacy software that is too brittle for realizing transformation goals.

Building upon the Issues and Strategic Zones of work, product leaders should start preparing their solutions repository with business cases from product teams and future capabilities they know they need and have higher confidence in delivering.

Level-three maturity allows these leaders to starting enhancing their system capabilities to new architectural patterns that are not in use yet, but often needed.

CASE STUDY: FOREIGN BANKING ORGANIZATION (FBO)

The head of product for a foreign banking organization (FBO) was tasked with trying to modernize a large existing mainframe—a level-two maturity objective. The organization's ultimate objective, however, was to gain a single view of the customer because they were not leveraging all their customers' needs, wants, and asks across multiple business lines. One of the critical stakeholders said, "If someone asks how many customers we have, we don't have an answer." There was no single way of identifying a customer, nor even a definition of what a customer is because there have been so many variations on the mainframe. They wanted to streamline and obtain an understanding, from a management standpoint, of who their customers are, what they do with them, and what kind of services the organization could additionally provide to those customers.

To achieve a level two maturity, first we created a business capabilities map. If we'd had to do a full business context diagram for the organization, it would have taken months. However, because the customer is a pervasive entity within the organization, we were able to stay at level one, mapping all the first-level capabilities they perform within retail banking, investment banking, and wealth management in Asia, Europe, and other geographies in which they operate. Then we looked at which of those capabilities directly touched a customer. We highlighted those at a level two in the business context—and then we moved on to the enterprise architecture system context, attempting to find all the applications where customer data was stored.

People, as we know, are reluctant to share information and sometimes honestly do not have answers, and nobody knew exactly where customer data existed—because, as we were often told, "Customer data exists everywhere." That might be true, but it had yet to be determined; thus, we had to achieve a level of maturity in institutional knowledge. We were able to filter out irrelevant applications, finally determining all the systems that needed deeper investigation.

We were able to answer many of the questions management asked, but the answers were incomplete. In level two, we saw that they didn't have proper data management, so they needed to re-architecture. In the level-three rationalization, we could tell them what needed to change, who needed to change it, and how long that process would take. We created a heat map and determined that they needed to obtain a master data management system and move all this data into a modern banking customer data platform and develop an integration plan. Once that was completed, this capability was once again properly aligned to the business strategy for this organization.

Through these initial two principles, we have observed that developing actionable business and systems capabilities maps will allow product leadership teams to measure business impact and customer experience, instead of focusing solely on technology KPIs or just chasing the next evolution of architectures and patterns. Next, we will look at data as the lifeblood that brings these capability maps' blueprint to life in Principle 3.

KEY TAKEAWAYS

1. A documented System Capabilities Map, overlaying Business Capabilities, will uncover institutional knowledge and insights that are never discussed at a leadership level.

2. Enterprise Architecture is a key ally in large organizations; leverage the practice to get ahead of rebuilding with the right approved tools.

3. Ability to sunset legacy platforms to support strategic business requirements is the first step on the digital business transformation journey.

PRINCIPLE 3

Use Formal Data Management Practices to Derive Economic Benefit for Customers and Business.

"Data is a precious thing and will last longer than the systems themselves."

—TIM BERNERS-LEE

Netflix, Airbnb, Amazon, and Uber have pioneered and practically proven that the value their platforms provide is in the data, showing that data is fundamental, especially when matched with frictionless transactions and the best possible customer experience.

The term "data" here refers to the information these platforms source from customers themselves, as well as external information from third parties, and the transactions generated from within the platform, not just orders, but reviews, quality, and timeliness of information. These companies were able to maximize the use of this information to facilitate a transaction that is seamless and rich with customer experience, whether for finding a ride or renting a property for the summer.

In contrast, large organizations—particularly financial services companies—have always dealt with data (market data, personal savings and expenses patterns, liquidity or number of tradeable products, research data) as their fundamental value of information, but that power has not been matched by great customer experience.

New entrants to the industry, on the other hand, focused on great customer experiences powered by data and simplifying the buying process, called frictionless transactions—for example, opening a bank account for savings, buying a stock, a fund, investing for retirement savings, etc. Robinhood's "Top Stock Lists" and "People also own" features, Goldman Sachs' "Marcus Insights" tool, and Betterment's "All-in-one" dashboard are all fueled by data. Imagine what large organizations could achieve if they matched similar customer experience using the vast power of data they already hold!

THE IMPORTANCE OF DATA

This principle focuses on how to evolve large organizations toward a data-centric product vision with focus on customer experience.

The development of data-centric products—not to be confused with data-*driven* product management (which is the process of being focused on making data-based decisions when managing the product lifecycle)—is a fundamentally complex problem for large organizations. The key is to develop product-oriented data management practices, which will be our focus here.

Few organizations prioritize having formal data management practices, but as the above platform examples demonstrate, valuable insights can be extracted from the data generated by an organization, which can then be used to enhance the customer experience.

According to EDM Council's Global Benchmark Survey of more than three hundred large organizations across thirty-five countries, formally managing data across the organization is less than a three-year-old endeavor for more than two-thirds of the respondents.[10] While the majority of those that started formal data management practices did so to support defensive activities such as government regulations, risk management, and other back office activities, very few, if any, focus on offensive strategies such as improving customer experience and delivering value to the customers.

Data is the lifeblood of an organization—but it has become both an asset and a liability since the financial crisis of 2008. Regulators and government bodies have enacted risk management and privacy restrictions on how data can be used, which is a large liability for product leadership teams to manage. The California Consumer Protection Act, or CCPA, is one example of giving consumers control over how their data is collected and used. (A similar regulation in Europe is the General Data Protection Regulation, or GDPR.)

However, with the advent of so many technological advancements, data is also becoming a first-class asset as a currency that can be used to improve products and enrich customer experience. Data can be used for commercial purposes, where the data itself is a product (as seen with Bloomberg), in advertising (Google), to better understand customers, in developing hypotheses about product features, or as information of value, sent to service providers to develop relationships that are focused on improving customer experience. The ultimate use of data, however, is to gain insights, which enable companies to provide more value to customers and enrich customer experience. In the past, data was relegated solely as a domain of technology teams, in data warehouses with occasional MIS reports; now the tables are turned and data has become a business-driven and technology-supported function.

In fact, it is not hyperbole to say that data is the key to an organization's survival and that an organization's longevity can be correlated to the health of its data. Newer companies currently emerge with an innate focus on constantly considering the data that they are generating and how it can be used to create value, but this process is not yet part of the muscle memory of large enterprise organizations. For example, Robinhood, a stock trading platform, leverages transactional data (similar to Amazon) by showing stocks that are of interest to other users, most popular stocks, other stocks owned by people who own a specific stock, etc. (although it can be argued that this process is gamified, for better or worse).

To put data to effective use, product leadership teams of those organizations need to know every piece of data used to build their products, to service their customers, and to market those products to the aforementioned customers. Whether a customer invokes the right to be forgotten or the organization needs to monetize data without violating local and international laws, the product leaders must be aware of the restrictions and rules surrounding this topic. Thus, product leadership teams need to find balance between that asset and liability.

Data is a broad umbrella term that comprises areas from technology to architecture to privacy regulations and more. In fact, a State Street Growth Readiness Survey found that, after cyber security, analytics is where the growth is—and a solid foundation of data management is a mandatory pillar for strong analytics.[11] Successful product leadership teams need to know certain aspects of the kind of data they, their internal partners, and their service providers collect and manage.

This principle will address the mindset around data, as viewed through these areas of product focus, called metadata or data *about* data:

- *How the data is created, including all areas of sourcing: provided by customers, vendors, and partners, and intrinsically created through the usage of the product (web analytics, customer transactions)*

- *How that data is used, owned, and how the content of the data (specific information, such as names, addresses, preferences) is managed within the organization*

- *How it impacts a product roadmap and which platform capabilities are foundational to use data insights within a product*

In this principle, we will focus on the principle that provides product orientation to data management. To put it more succinctly, every product leader needs to know the state of their data, and they need to take that data into consideration when building products.

THE IMPORTANCE OF DATA MANAGEMENT

Data is not just a byproduct of the transactions performed or the capabilities offered by an organization; if managed correctly, data can generate significant value for that firm and become a key differentiator. Teams able to advance the maturity of their data management practices can glean insights for the organization or for their customers from the mountain of data created or enhanced, which can then be used to drive business decisions and deliver differentiated user experiences such as personalization, recommendation engines, and predictive actions.

This doesn't happen by accident, however; offering either the internal analytics or external enhanced experiences can only be accomplished by having effective data management around the data generated by the product. McKinsey defines this process as "AtV" (Analytics to Value) for the enterprise.[12] Product leaders at large enterprises often do not own all the necessary data needed for the inner workings on the product. Ownership refers to being able to own the lifecycle from end to end, i.e., sourcing from the original provider, including the customer or partner, through destruction of the data on-premises. Given that the ownership of customer data, market data, and vendor data is "owned" by multiple functions—i.e., sales, marketing, business, technology, and operations—data is a federated entity. Federation is the source of many problems within the organization. To manage this federated process, product leaders need to institute formal processes into their teams so that they can create a robust vision for leveraging data and so the engineering organization can derive momentum by building products and features driven by the rich data collected. Products are differentiated not just by the capability (which may be provided by other similar products) but by the *quality* of the experience enhanced by data-centric insights.

Such experiences can only be delivered when product teams harness the power of their large data sets. Once data management practices reach an advanced stage of maturity, new, as-yet-unimagined features can be released leveraging the product data.

In addition to building on data, product teams may also be able to monetize data sets by, for example, generating and leveraging aggregated insights from their data or building machine learning processes around their data to enhance the product. As a more specific example, Uber collects significant data around traffic patterns. While they currently provide that data to municipalities for free, hypothetically they could choose to sell it to a construction company that wants to make a better pitch for an expanded highway or a different lane configuration. As a real-life example, while controversial, Robinhood routes data to brokerage firms that pay for order flow.[13]

Additionally, a product leadership team can confirm that

their organization is properly capturing data so that they know not only where the data is stored and how it is cataloged, but also that it's vetted and validated, removing potential reputational risks to the company. If, on the other hand, the data has not been captured appropriately—or the appropriate data hasn't been captured—it may reflect incorrect information when used in a product, and ultimately, the reputation of the product falls on product leadership, so it is in their best interest to quality check the information flow.

When using data, product leadership teams must follow the proper processes to ensure that data is fit for purpose—whatever the purpose is determined to be. Any product leader with a large data set responsibility, whether they are using that data defensively (for compliance with regulations) or on offense (to enhance products with the data they already have), needs to achieve a level of maturity before putting it to use, which will be described in a later section of this principle.

To return to the example of the California Consumer Protection Act, a defensive example of knowing where data sits within the product portfolio, a customer from California logging in to a website has the right to click a button that says, "Forget about me," which then requires the organization to remove all of that customer's data or stop using it—whether it's used and stored in one product or across many different products. A business that violates the CCPA,[14] by not providing that option and following that process, will be fined 4 percent of their revenue.

Even businesses that do not hold customer data face downside risk. The CTO of a company selling kitchen management and catering software deals solely in business-to-business (B2B) data, rather than business-to-consumer (B2C), but he still must be aware of regulatory challenges such as customers' personally identifiable information (PII) data, credit card data, etc. The regulatory risk varies by industry, but whatever regulations the organization faces, the product leadership team must first look to the data and confirm their data management follows those regulations.

It should be obvious but is worth stating that knowing the customer today is no longer limited to simple demographics; product teams must also have a deep understanding of their habits and preferences. A wealth management firm rolled out an improved tool to help their customers with tax planning. After it was released, however, the data showed that customers not only stopped using the tool, but they also failed to move forward with calling service representatives for help. The tool was so complicated that the customers simply abandoned it. A product leadership team analyzing the data collected around that tool would be able to see that trend and determine the best action to take.

Product leaders also need to understand the data lineage, the full supply chain of data that is being used, before introducing a new feature leveraging this data. When product leaders can clearly see where the organization is capturing information, ownership of data along the way, why that information is captured, and how it is used, what it can be used for, etc., this information enables product leaders to explain to their team the new change to be implemented and how it affects the product so that the data can be properly used and stored to be in compliance with governance policies and procedures. When product leaders have this information in a structured format, they can quickly perform impact analysis, particularly valuable as larger changes occur in the industry, or a new product rollout impacts assessment.

Without a structure that provides useful metadata in an organized fashion, the product leadership team may not know where to start implementing the change or could become mired in searching for information in the wrong places, leading to a significant delay in implementation rather than being able to move quickly and effectively. McKinsey estimated that this time could be up to 74 percent of employee time.[15]

COMMON CHALLENGES IN IMPLEMENTING THIS PRINCIPLE

The EDM Council's Benchmark Survey showed that formal data management practices are still relatively new behavior for organizations.[16] The concept of having a chief data officer, or CDO, at the executive table only began to emerge around 2015. Organizational behavior needs to become oriented around becoming data-first organizations—but product leadership teams may experience some challenges on that journey.

Ownership of data is a new and evolving concept within large organizations.

By far the largest stumbling block to product leadership teams taking full ownership of the data comes down to time and dedicated attention. Organizations that dedicate effort to establish a baseline understanding of the data do so by allocating the time and energy to building out an information map as a guiding tool, which we will be breaking down under our "Data Operating Model" section later in this principle.

Busy product leaders at enterprise organizations don't call in experts to perform this data management assessment because they are unaware that they need to do it or are *unable* to do it themselves; they do so because they simply don't have the time. They are constrained by other responsibilities, and they need this assessment completed more expediently than they themselves can commit to.

Even product leaders who can schedule the time and dedicate a team to creating the first business information map, at businesses where the product leadership team works toward taking ownership of the data, may still find it unclear *how* exactly to go about doing so. The shift from technology ownership of data to product ownership has caused an emergence of new approaches in data management and governance.

Complicating matters further, shared data ownership for some parts of the enterprise data makes it difficult to assign ownership responsibility to large, shared, and regulated data entities, such as "Customer" data or "Market" data. CDOs are provided with the mandate to resolve these challenges, and the success of it will be driven by established governance models. Product ownership should be encouraged to work with CDOs to create proper governance on shared data sets.

Budgeting for data technology requires critical buy-in.

On the technology front, previously available technologies for utilizing massive data sets were cost-prohibitive, thus only a small fraction of companies could take advantage of those technologies. Today, with the decentralization of infrastructure ownership via the cloud, there are more options for all sized firms to take advantage of data engineering and science capabilities, provided they follow the basic principles that will be outlined in this principle, to catalog, track, and store their data correctly. However, because this is still a relatively new concept, and because the immediate return on investment for these activities may be unclear, product leadership teams may encounter difficulty in articulating its value, which can make it difficult to get business buy-in. It is rare, when establishing data management practices and principles, to deliver near-term value that can be monetized; many of these are proactive measures that unlock insights over time.

Therefore, to win business investment, data management must be embedded in the context of an existing business engagement. If, for example, one of a leader's initiatives is to introduce new capabilities into a product, then while building out those capabilities, the leader can follow this principle and advance the maturity within the subject areas that are affected by that capability. It is easier to work in the context of a business, risk, and regulatory feature and obtain approval, rather than making standalone data management investments. Too many times, large organizations follow a top-down "if you build it, they will come" approach and fail when it comes to data management.

To look at a specific example, a CDO at a wealth management firm did not want to implement a standalone data management initiative when maturing how the organization manages data for product improvements. This CDO also felt that cataloging and cleaning up data were going to be an upstream sell because those activities provide no immediate business value. Thus, this client rolled data management into part of a larger program they were building to address a particular business function supporting high net worth individuals. They examined the data they included as part of this new program and followed the process detailed in this principle, making that process a repeatable, foundational aspect of existing work. As they perform more and more of these transformation initiatives, they will be able to scale that process so the business can see the value of doing this work. First, the client showcases some wins with data, and then they seek to make data management more of an institutional process.

Many organizations invest in a data management platform before having a solid understanding of the product data ecosystem.

An additional challenge comes in attempting to progress out of sequence. Organizations may be quick to make heavy investments in large data management platforms without first achieving maturity in the other foundational aspects discussed in the next section. Because they haven't solved the precursors that enable utilizing those investments more efficiently, they do not achieve maximum value from those platforms.

We've encountered more than one enterprise client with teams trying to build large data warehouses, just for the sake of it, with no clarity as to who's going to use it or what value it will provide (again, the strategy of "if you build it, they will come"). We've also seen clients with multiple partnership agreements signed with multiple AI firms with no tangible benefits; without properly governed and high-quality data, those data centers and tools end up unused or worse, eroding the trust of the business.

DATA OPERATING MODEL

Before progressing through the maturity model, the prerequisite step is to identify a data operating model, meaning, how are the business and product engineering teams organized to operate daily pertaining to management of data sets they own?

Establishing a minimal operating model, or what the industry calls a data governance model, is imperative because product leadership teams need buy-in from business and technology stakeholders in order to fund this work.

This step determines the kind of ownership that can be assigned to this work so that the information can be captured and used for answering various questions, such as:

- *What data do we have as an organization?*
- *How is that data generated?*
- *Do we have good quality data?*
- *Do we have any privacy or ethical requirements for using this data?*
- *Do we store any sensitive customer information?*
- *Is any of this data leaving our organization? If so, why?*
- *Who is the first person to call if a regulator from a government agency says, "We found that customer personal data is being misused from your organization"?*

Most product leaders have similar questions, but they don't know the starting point or process to find the answers to those questions. Having an approved enterprise data governance model allows this work to be completed successfully. Many product leaders may rely on CDOs to do this work for them because those product leaders are not able to do it all on their own.

Assuming a fit-for-purpose data governance model exists, which is not within the scope of this book (for reference, multiple bodies of knowledge exist in the data management industry supported by various consortiums, such as EDM, DAMA), product leaders need to understand the governance model to answer the previously listed questions for the product or portfolio of products they manage. The product leadership team can then begin to progress through the maturity model development we outline here.

An established enterprise data governance model will allow product leaders to be set up for success. In large organizations, as data management programs mature, this operating model is becoming part of the software development lifecycle, and product leaders should use it to their advantage and establish a minimal model prior to maturing on this data-centric products principle.

As an example, when developing a new feature for the product, the data governance process will call out the need to document the data dictionary of the feature, validate that regulatory controlled data is managed properly, and the lineage is documented accurately. These lifecycle rules from the governance should be leveraged by a product leadership team to improve the state of their data sets. It is possible that some of the work might be already done by centralized data management teams.

MATURITY MODEL

Maturity in this principle occurs across three different axes:

- **Axis 1:** *Create a consolidated view of the critical data entities through a data catalog, within those subject areas for which the product leadership team has oversight and governance.*

- **Axis 2:** *Create a data lineage for data being generated, consumed, and enriched to drive insights. Build a process for teams to quickly replicate this for the next funded capability.*

- **Axis 3:** *Analytical data is enriched and enhanced for differentiated data-centric products and a data monetization strategy.*

A product leader should aim to achieve the following levels of maturity with their data management practices within their product portfolio:

	Level One	Level Two	Level Three
Data Catalog Axis	A catalog of data domains (types of data) for the product is formally established and terminology is created for daily use.	Most critical attributes of data within the catalog are formalized and ownership is established for business fit.	A formal change management process is set up to curate the catalog and is distributed regularly to product engineering teams.
Data Lineage and Quality Axis	Ability to visualize the flow of data domains across the product teams, operations, systems, and partners exists.	Lineage of data and data quality ownership is established.	Data quality reporting and remediation is formally part of the product lifecycle.
Data Platform Support Axis	Product engineering teams have formal processes to review the data catalog and critical data elements for enhancing product features.	Data and analytical platforms are built to be able to define and realize use cases.	Product engineering teams have the ability to leverage and build data-centric features into the product to differentiate themselves in the market.

- *Level 1: A high-level understanding of product leadership's data domain; end to end flow of data, processes, and owners; and establishing a process to discuss usage of data in product core features. Establish a technical program to integrate the data for insight generation.*

- **Level 2:** *Dig deep into the core attributes of the catalog, how each of these critical data elements (CDEs, as they are known in the industry) are flowing across system boundaries and ownership of quality. Identify specific features to be built from Level 1 use cases.*

- **Level 3:** *Establish effective change management across all the axes of the data operating model. Implement data-driven features derived from the analytical process created leveraging the data technology platform(s).*

The rest of this section will look at this maturity model in greater depth.

Level One

DATA CATALOG

Similar to the business capability map generated in Principle 1 and the systems capability map from Principle 2, level one of this principle leads to the creation of a business information map, which organizes the information the business collects— the nouns and verbs that make up data.

A data catalog predominantly contains verbs, features the business offers that a customer can *do*: update their profile, give a gift card, call a taxi, etc. The architecture map illustrates the systems that enable those verbs: for example, three systems that deal with taxi booking, two systems that deliver gift cards. The data on a business information map, however, is both. The data will track nouns, like a customer, a driver, a taxi, a receipt, all of which are represented in a database as one or more entity types. That database also captures verbs, the transactions performed by nouns over their history with the company: the trips a customer has taken or a driver has driven. All of that data—the verbs and nouns—can be logged, mapped, and analyzed as appropriate.

Here is a simplified example of what this could look like for Amazon's book business:

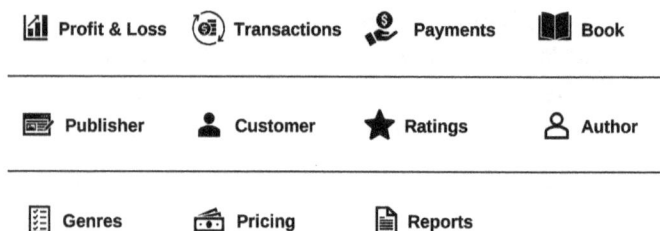

Profit & Loss	Transactions	Payments	Book
Publisher	Customer	Ratings	Author
Genres	Pricing	Reports	

Figure 10: A high-level data domain of Amazon's book business

Once the product leader has mapped those applicable nouns and verbs, the next task is to organize them into groups. That allows the product leader to both have a method to the madness and to show product teams the areas where they may want to focus. If they need to understand more about customer profile attributes, for example, they can look to that section of the business information map.

In the Amazon example, a product leader may want to catalog all the products they sell. In that case, they will organize the boxes labeled Book, Author, Genres, and Publisher into one bucket and assign an owner who will obtain that information. The product team would define the "taxonomy" or classification of entities that define a group—in this case, the product catalog for a book could be extended to other product lines in the future if, for example, electronics were added to the portfolio. The product leadership team would need to understand the hierarchy of information available in the product.

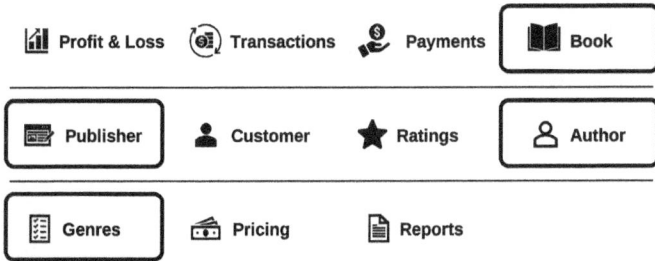

Profit & Loss	Transactions	Payments	**Book**
Publisher	Customer	Ratings	**Author**
Genres	Pricing	Reports	

Figure 11: Bookstore Product Catalog Data Domain (highlighted)

The data catalog should identify all the nouns and verbs, their groups, and their ownership within the product organization. For example, the head of sales may own customer data, while the head of operations owns all transactional data. An owner can be assigned to each group, who can then align the buckets in terms of how the company does business. Different organizations will organize their boxes differently—even if the boxes themselves are the same. The only way to do that is to look at how the company is organized, not necessarily with regards to people but in terms of product lines, business, or geographies. Ultimately, this isn't just to know who owns what; the value for the product leadership team is in determining the best way to manage this information so that they can make decisions.

To begin, rather than boiling the ocean to map all products, the product leadership team may choose to create this map for the line of business most in need of improvement or one for which funding has already been approved. The very first step is to implement data-driven practices in the organization and create a catalog of data.

For example, assume that a financial services firm implements a new program and wants to evaluate the product portfolio to see where they collect cash balances from all different types of customers—retail clients, institutional clients, etc. At level one, they can begin documenting all the places where this new domain "cash balances" occurs, perhaps across multiple geographies and in multiple currencies.

After the first iteration of this level, the product leadership team will have taken stock of the metadata, put it in a proper taxonomy to organize it, and can potentially use it to report back to the business about the data that may be useful for regulatory purposes, to build data-driven products, or for the organization's understanding of how to leverage this data. From there, level one can be repeated with another line of business, per the first axis previously discussed.

The biggest problem we see at this level is with the terminology. The way in which one person defines the term "customer" may not be the same for someone else in another area. For a salesperson, the customer is the name, address, and phone number, whereas for a trader, it's his account number, background check, and license information. The nouns are very contextual, so people may fight or get into semantic discussions over how they are defined. When that is the case, we recommend setting aside the differences for now, simply entering different boxes for customers, and resolving the domain boundaries later through change management as the organization matures to level three. It's more important to move forward, even with multiple customer entity types, than to do nothing.

DATA LINEAGE AND QUALITY

Once the data is captured and owners are assigned, product leadership teams can consider where this data is coming from—the supply chain or lineage of the data. At this level of maturity, product leadership should focus on domain lineage, which is the understanding of how data entities (Customer, Product) move from system to system, shared across lines of business, who touches and modifies them, and whether they contain privacy fields that are sensitive for use.

At its most basic, this lineage seeks to answer, "Who uses what?" A claims system may use customer data, for example, to create a new claim, pay out a claim, and to receive payments. The product leadership team has the job of connecting the dots between the system and the business with regard to how those nouns and verbs are used within the product portfolio.

Other questions product leaders can answer at this level include:

- *Where does the company get the data? What is its source? Does the organization create the data, or does somebody else create it? Does the company buy the data and then use it?*
- *What journey does the data take within business context, systems context, in the universe of products?*
- *Who are all the people who use the data?*
- *How and why do they use it?*
- *Who is the contact for this information so the product leader knows who to talk to?*

This creates the data supply chain or lineage—a document that shows, for each noun and verb, how that data flows through the entire organization, through systems, people, processes, etc., from end to end, and what happens to that information throughout the business capability map and throughout the system context.

A customer may be one entity in the level-one business information model. In the lineage, customer data may come from a customer relationship management system, where it's first created and stored, then it goes to a trading platform, then through a different accounting system, then to invoicing, on to analytics, and finally to a report.

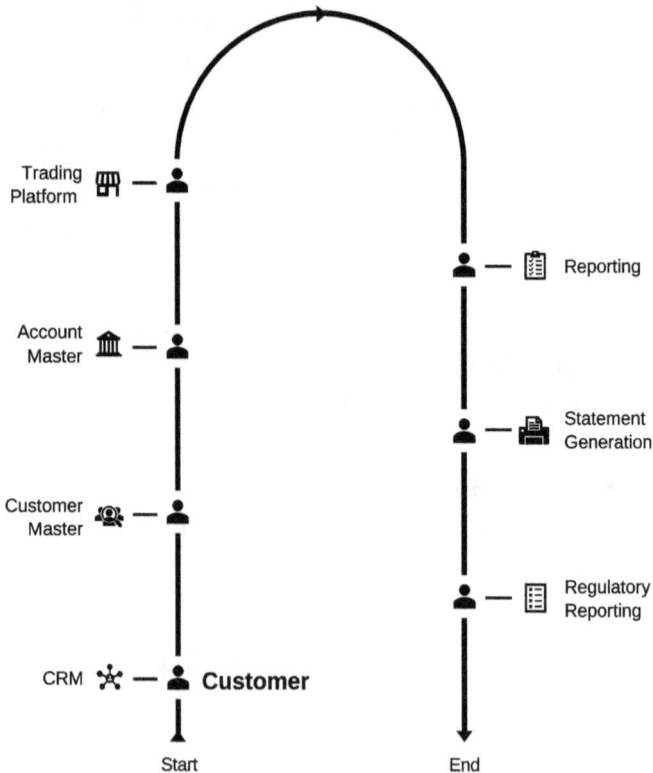

Figure 12: An example of customer data lineage in a retail financial services organization

This lineage allows the product leadership team to answer, "What do we do with customer data in this organization?" Then, as a hypothetical example, if an invoicing system has to be replaced, the product leader can easily see whether that system handles customer data and the impact of the new invoicing system to the business information model, mapped to the business capability and system architecture. This is useful for understanding what's happening to data within the organization and allows the product leadership team to answer more questions at this level compared with the previous.

DATA PLATFORM SUPPORT

The final axis of interest is developing data-centric uses for the product or product portfolio. For example, for a bond trading platform, this could be showing other available bonds with similar coupon rate(s), the total volume of bonds traded relative to today, latest bond news, etc. This will make the customer experience better and more well informed and will develop stickiness to the product.

One of the core tenets of level one is to start planning a data technology and analytics platform so, as the product team matures, the insights generated from vast amounts of data can be sourced by the organization and put to good use.

Level Two

DATA CATALOG

The second level of maturity focuses on specific CDEs that are stored, how they move within the organization's capabilities and technology, the quality associated to those critical data elements, and what specific features will be developed using these attributes.

A CDE is one of the important attributes of an entity (for example, a Customer Entity has Customer ID, First Name, Last Name, Address, Social Security Number (SSN), Marital Status, Zip Code, County, etc.). This level focuses on determining critical elements for each of the nouns and verbs from level one. For example, if the noun is "customer," what are the twenty critical data elements that make up "customer," and are those attributes equally necessary for the verbs "trade" or "claim"? This level of maturity allows the product leadership team to understand which business entities have critical sensitive data, what we call PII data, as well as business critical data for operational and analytical use in the future.

A specific business line, say a trading group, might not consider county and marital status useful data elements for their business purposes, but the rest are considered important to run the business effectively, so those remaining elements become CDEs. CDEs are completely product specific, however, so a different wealth management or marketing group within the organization may find the marital status of the customer useful to drive campaigns.

It is very important for product leadership to have a good handle on CDEs and the implications of having bad quality CDEs. For example, if a customer's social security number is not captured correctly, it may not be possible to conduct a background check on the customer before allowing money to move, in line with local regulations, whereas if the marital status is not captured, it may have no bearing on the product. Data quality on CDEs has the potential to cause reputational damage to the product or the organization at large. In some

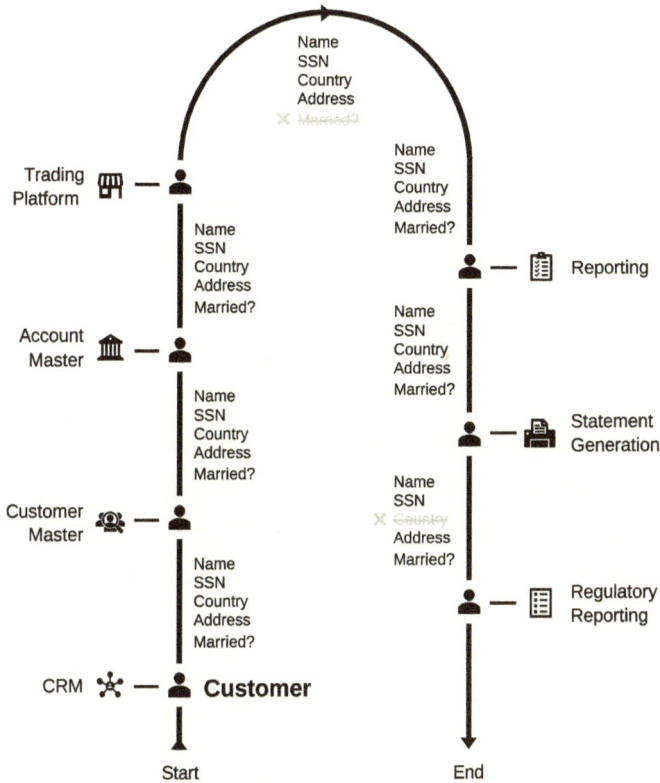

Figure 13: Marital status and county removed from tracking

cases, regulators define rules to force organizations to maintain data quality on these CDEs; BCBS 239 is an example of this in the financial services industry.

The adaptive and dynamic understanding of this information model and how that data is used drives the maturity of the operating model and, to be clear, the maturity of the product leadership in leveraging their own data. From a governance standpoint, this level also means having a playbook ready to answer analytical, compliance, or regulatory questions, such as, "What do you do with customer data in your organization?"

Here, the product leadership team will want to go into greater detail about this domain, from a business standpoint:

- *What is the critical data?*
- *How much data is shared with someone else?*
- *How much data is related to consumer privacy?*
- *How much is related to access control?*
- *What is related to data production? How much is related to other elements?*

From the technology side, they will want to know:

- *What are the systems mapping into the enterprise architecture where this data is stored?*
- *What do I need to do to get new data into the system, or what do I need to do to improve use of this data by some systems?*

Where to go deeper is a factor of time, effort, and budget. Many heads of product focus on those kinds of entities that have potential for PII data because they have compliance teams behind them to assist with organization of this data. Level three can also be used to address customer worries, regulatory compliance, and/or the introduction of new products or features. However, if any element of data quality threatens reputational risk, that should be the primary focus.

DATA LINEAGE AND QUALITY

In addition to CDE inventory by domain, this level also focuses on creating lineage for these CDEs. Data lineage is a systematic documentation of the supply chain of data coming into the product, being used within the systems and services provided by the product, and ultimately shared with customers or external service providers of the product. Product leadership teams can effectively use the data to identify features to provide to the customer in a small, contained fashion to differentiate the customer experience. This can only be done, however, if the product leader knows that they have good data, knows how that data has evolved over time, has an awareness of the kinds of issues customers are experiencing, and can then work with the product team to enrich the customer experience.

If the customer data originates from a customer relationship management (CRM) system, there needs to be good controls in place to make sure either the CRM captures all of the CDEs, or Line of Business (LoB) has policies and procedures in place to reject customers without critical information (such as SSN or Address, from the previous example).

A common problem product leaders encounter at this level is attempting to create this lineage without having first looked at systems context and enterprise architecture. If Principle 1 or 2 is missing, the product leader won't have full context for the lineage of Principle 3 information. Depending on the question being answered, the deeper they have progressed on the first two principles, the richer the information gathered here—but that doesn't prevent strategic product leaders from gleaning valuable information from this lineage, even having fulfilled only level one on Principles 1 and 2.

Data quality is another important aspect of good data-centric product management. How is data quality defined? According to the EDM Council's definition, data quality is information that is fit-for-purpose for business use. It simply meets the needs of defined business processes and outcomes.

In level two, we are going to start by understanding the quality of existing data and reporting on that data quality. Increasing visibility into the quality of data is a good first step to fixing it.

Product owners should define which quality rules (e.g., missing customer's zip code or invalid zip code will cause return mailings and sensitive data to be mishandled) are going to impact a product's business processes or negatively impact the customer experience. In level two, we track the quality rules and percentage of data that is impacting the customer experience negatively.

DATA PLATFORM SUPPORT

Finally, from this capability's axis point of view, the product leadership team should start planning specific features out of the use cases discussed in level one. For example, if a financial services firm has the data set for cash balances as one of the CDE attributes for a customer, what can the product team do with that data to show incremental product improvements and business benefits? Perhaps they can proactively warn customers that, based on their average monthly expenses tracked over the past six years, it appears that next month they are at risk of having a cash shortfall. During level-two maturity, product leaders will start setting up the data platform to support such capabilities. Whereas previously all that was shown was the balance, now insightful information is being provided to the client, who then knows that, unless they move money around, they are going to run into this cash shortfall situation, which may affect their own operational needs. That customer can now also look ahead to see if they have another potential cash shortfall the following month—and they are not likely to go to another company for cash management when the current company helps them with more proactive measures for managing their cash.

Level Three

Level three then builds upon the understanding around the use of data in building new products, called change management.

DATA CATALOG

Change management describes the process used to move an enterprise platform to a new capability level by adding features. This principle is about data, so we are referring to data-centric features of the product, not generally described by traditional business users or operations experts. This is how the product gets differentiated in the market.

What does supporting a new data-centric product feature mean? As part of maturity building up to this level, we have answered questions, such as which systems have needed critical information, what is the supply chain of that information, and do specific systems have critical information needed to enhance the product? A data-centric product feature means developing a product feature that is powered exclusively by data available within the platform or sourced from the participants and stakeholders, not a transaction or business logic.

In level three, product leaders will show the business how customer experience can be enriched by following this process of collecting data, tagging data, creating taxonomy, and enriching product features—showing that it is a repeatable process and demonstrating the business and customer value in it, thus securing business buy-in to become a data-centric product company—the leadership team is now going to build a foundational platform to house all of their data. That data platform enables product teams to self-serve in defining and building data-driven features at scale (for example, insights for the platform users, search capabilities, recommendation engines, monetizing data, and AI/chatbots). These data-centric features will become a roadmap on their own, and the organization can focus on differentiated experiences, not just managing transactions.

In a financial services example, a brokerage platform will know how much cash the customer has in their account, called cash balance or "buying power." In many organizations, this balance just sits there. A data-driven feature for the customer could be a chat with a bot to see their cash balances or a determination based on their previous history that they should potentially be investing cash somewhere else, with a nudge for the customer to put money to better use.

The primary opportunity and challenge are that the more in-depth heads of product go with their data, the more information they have about the entire supply chain. If a company captures ten attributes from a customer when they register on a site, then that company bears the onus of explaining what happens to all ten attributes across the entire organization, which quickly multiplies the amount of work the product leadership team needs to do. Thus, critical data elements should be limited to those that are truly crucial—as the name suggests, focus on what is *critical*. (But product leaders should also bear in mind that other non-critical attributes will still exist in master repositories, whether they have good quality or not, because what might not be critical to one product may be critical for someone else, as explained in an earlier level with regard to marital status.)

DATA LINEAGE AND QUALITY

When it comes to lineage and quality, the remaining focus is on improving data quality, based on the reporting from level two.

In large enterprises, data quality improvements require extensive coordination based on data lineage understanding. The best place to improve the data quality is at the source; all others are just temporary solutions. Knowing lineage will greatly help, and the data operating model will help establish a feedback loop to measure progress.

One critical aspect of data quality remediation is quantifying the business impact of a data quality issue. For example, what is the impact to business when address or tax identification fields are inaccurate? What is the impact when address changes are not revised in the customer master? Some answers may be: mail marketing waste, possibility of sending statements to the wrong address, or incorrect tax reporting.

Once the proper case is made to remediate, these become engineering or data engineering tasks for cleanup, and the data quality reporting process will measure progress against such requests.

DATA PLATFORM SUPPORT

During level three maturity, the product teams can take advantage of the data platforms that were built in earlier maturity levels to build sticky and differentiating features for their customers. We will pick up on the cash balance example we discussed during level two that was developed on the data platform.

With a robust, trusted historical cash balance at hand, the product team can add features such as forecasting cash balance, alerting on historically low cash balance, giving advice on how to move the money around, stopping a scheduled spend to keep cash balance at a healthy level, or even suggesting new products, such as Line of Credit.

The final component in level three is requesting real time feedback on data-centric features. This is the critical flow-back function into the change management process.

Unlike traditional application features, such as trading, buying a policy, or sending a balance notification via SMS, not all data-centric feature impacts are known in detail beforehand. For example, while a recommendation engine might generate recommendations based on data available within the platform, product leadership may not yet know what sort of patterns will develop over time. Thus, it is critical to solicit customer feedback on data-centric features, either through usage analytics or by talking to customers, to make sure they are iterated on along with new and upcoming features. For example, when a chatbot responds to a user query, the bot could ask if the response was appropriate, perhaps with a like button.

At this level of maturity, the product leaders are in sync with managing data, metadata, and quality associated with their products to be compliant with various regulations and enrich customer experience for better differentiation in the marketplace.

CASE STUDY: BOND TRADING PLATFORM

A new head of business for a bond trading firm inherited a third-party legacy solution and wanted to build a unique platform with a larger pool of bonds to trade from, intuitive features, and a data-centric differentiated offering.

One of the key performance metrics is to measure the percentage of maturing or upcoming debt the platform could automatically roll into new bonds. An example might be the rollover of a Certificate of Deposit (CD). Using transaction and holdings data generated by the platform, we designed and built a capability that automatically suggests which new similar bonds the trading platform should roll the cash into upon maturity.

This case study highlights the notion that data generated by the customers using the platform can build differentiated features at scale and become an insightful platform that can enrich customer experience.

CASE STUDY: INVESTMENT ADVISORY SOLUTION

An investment advisory firm that customizes trading and portfolios for affluent investors wanted to provide insight into investors' demographics, stocks, and fund trading habits with similar portfolio diversification interests.

The firm was able to pull in data associated with specific trades assigned to various portfolios with similar risk profiles. This data was then attached to market trading and holdings data across the entire company. The combined view enabled a broader perspective to the advisors and clients into the type of trades that they could execute to improve their risk/return.

This case study underscores that crowdsourced and market data can provide unique insights for targeted customers with similar interests. While there is always a risk of treading into the territory of gamification in this context, looking for patterns with guided advice can unlock otherwise hidden insights.

A robust data management discipline we discussed in this principle is imperative with differentiated cases such as above. Developing features such as these, without proper data ownership, governance, and quality controls discussed as part of this principle, could cause distrust or worse compliance issues for the organizations.

Developing data-centric features is a crucial differentiator for products in the marketplace. These features can create stickiness and facts-driven customer engagement. Now that we have discussed the core product, engineering, and data principles, we will discuss how to mobilize the teams the right way to achieve momentum across these areas.

KEY TAKEAWAYS

1. Data is an organization's key product differentiator and must be leveraged both for defensive (regulatory) and offensive (product) purposes.

2. A carefully drafted data operating model must be defined to manage data effectively, and a gradual maturity evolution will occur.

3. Data Cataloging, Lineage and Quality Analysis, and Data Platform development are the main axes for effective data management in product-centric organizations.

4. Product leaders should make it their objective to develop data-centric features and leverage insights from organizational data assets.

PRINCIPLE 4

Customer Needs Should Drive Your Agile Methodology, Not the Other Way Around.

"There is no such thing as a good or bad organizational structure; there are only appropriate or inappropriate ones."

—HAROLD KERZNER

The head of a product team sought our assistance in developing solutions for an issue that was inhibiting their ability to be responsive to end customers. After surveying internal business teams, we received feedback stating that requested features were not made available within an appropriate timeframe. Because the internal teams were organized by skill specialization (for example: web team, testing and release team, infrastructure team, deployment team), there was no consistent view of where the customer's holistic ask appeared in the lifecycle—thus requiring consultation with several different teams to determine when each portion of a feature might be released.

After an initial conversation, it became obvious that this particular product head did not have an effective Agile organization strategy in place to keep the information flow—all the way from the product manager down to the teams—consistent and locked in. This was a surprise to the leader because just a few months prior the team had internally gone through what was perceived to be a Spotify reorganization. Although this team seemed to have all the necessary ceremonies, beneath the surface the organizational structure was suboptimal.

It was discovered that the teams had implemented Spotify on paper but ended up with an inverted organizational construct— they ended up with a structure where product features were owned by multiple teams of specialists, instead of multidisciplinary squads owning features in their entirety. The internal customer consistently received conflicting information about the state of their project, and it felt like they were constantly conferencing with several different people for clarification. This was a case of a framework being "slapped on" a team without consideration of the team's true needs; leadership hadn't layered in the critical piece: the flow of work from the initial customer ask, through the team, and back out to the customer.

That level of dysfunction created an unhealthy measure of pressure in the team, which then manifested in other issues: People were unhappy. They did not feel a sense of pride in working for this team because it could never meet its objectives. Internal politics led to loss of funding and attrition. Externally, disappointed customers led to a loss of business.

This may be an extreme example, but it serves to highlight the way a team that is unable to deliver against its objectives, owing to an improperly applied framework, negatively impacts an entire organization.

THE IMPORTANCE OF ORGANIZATIONAL EFFECTIVENESS

Most people are familiar with the feeling of being part of a team that operates as a well-oiled machine: a sense of harmony amongst the team, an ease in working together and with their shared service teams, and the security of having a leader who has pulled together the right people to support the right products, using the right organizational frameworks. As a result, that leader can trust their teams to execute against the organization's vision, knowing that product delivery will be predictable and efficient.

But many, if not all, of those who have worked in large organizations are likely to have also experienced the opposite: being part of a team where nothing is smooth, an experience rife with confrontations and collisions. Perhaps that team is unable to meet objectives, or they are frequently rebuked, leading to feelings of stress and anxiety. Talented people from that team are likely to have other employment options and may choose to leave, leading to attrition for the organization. No matter how grandiose the end vision product leaders may have, if their teams are unable to fulfill their commitments, that vision will never be realized—and the leadership team may even find themselves suffering dire consequences, perhaps including the reduction of the team and termination of team members.

We frequently work with leaders who believe that implementing one of the many available Agile frameworks can help their teams become more productive and increase delivery efficiency. They may assume that doing so will help their teams become more in line with their technology upstart counterparts, who are by default structured this way, but the application of these frameworks as seen in large enterprises tends to be biased toward applying the more textbook implementation of the framework instead of understanding the philosophy and underlying principles that help achieve the intended outcomes.

Scaling Agile in a large organization has been found to be possible, but it's not easy. There are some misunderstandings about what scaling means and how many different methods can work for an enterprise-scale company. Each framework has its own pros and cons that need consideration before diving into any one approach headfirst. As illustrated in the State of Agile survey, a range of frameworks have been applied by many organizations, but a large number of those organizations have learned that the application of any framework, in and of itself, may not yield the market responsiveness that was initially sought after.

Incorrectly applied frameworks can cause unforeseen issues. Inefficiencies may be introduced into a delivery team because a percentage of that team's mental bandwidth will be

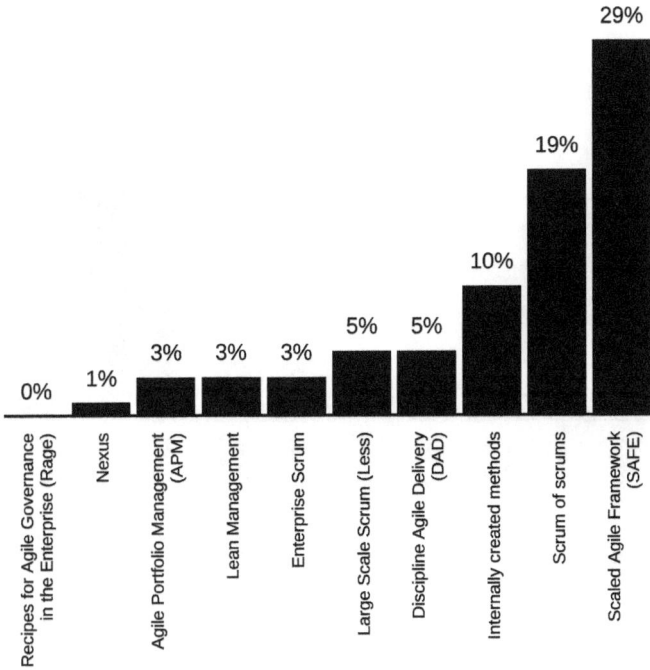

Figure 14: Scaled Agile framework adoption[17]

devoted solely to understanding how they are supposed to work, which takes time and mental focus away from performing the work itself. Implementing an Agile framework in a product team without a firm understanding of the delivery cycles and customer expectations will cause additional context switching and exacerbate existing delivery issues that the team faces.

Rather than beginning with textbook Agile implementation, using a framework as-is and expecting the desired outcomes to automatically follow, product leaders should begin by assessing the expected outcomes for their teams and then tailor which elements of Agile frameworks to use so the team can deliver in alignment with the overall mission of the

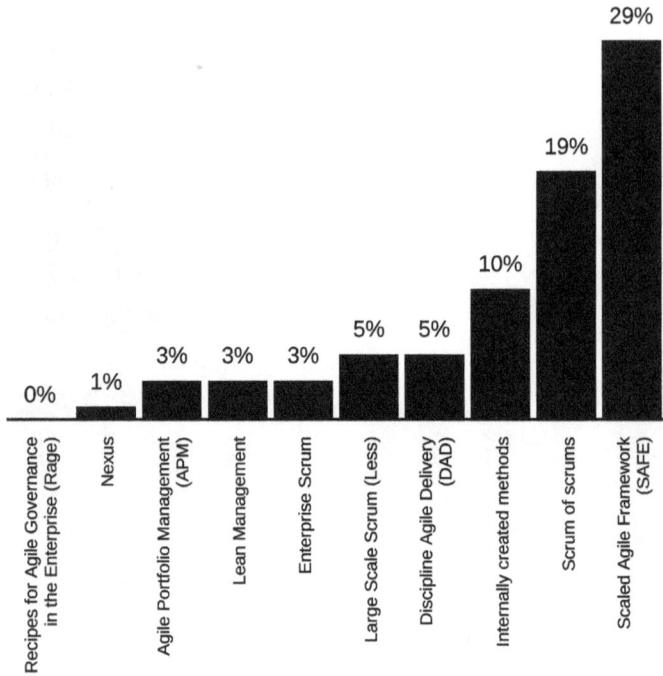

Figure 15: Illustrative differences across popular scaled Agile frameworks[18]

company. As product leadership teams embark on customizing the framework, they should constantly keep their customers' expectations in mind and use parts of any framework that help meet or exceed those expectations.

MATURITY MODEL

Because most teams are likely to have already adopted some version of an Agile framework, the product leader needs to first benchmark their current Agile state. Then, they need to have a practical understanding of the delivery dynamics of the product

A cPrime survey of customers found that a benchmark of customers surveyed between 2017 and 2020 demonstrated a shrinking adoption of frameworks and increasing demand by survey participants for flexibility within the frameworks that were selected.

Framework Adoption Trends

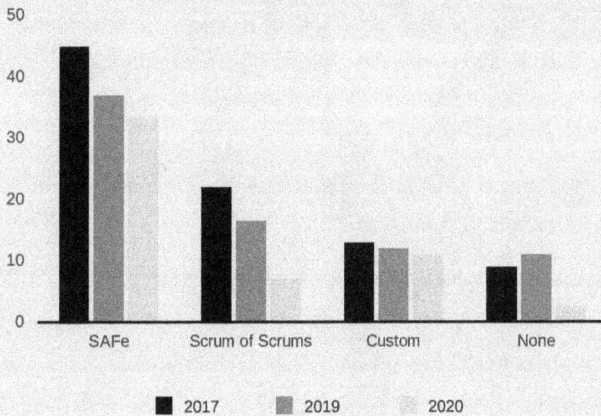

Figure 16: Agile framework adoption trends over time[19]

portfolio itself. What is an ideal delivery cadence? What is expected in the marketplace and/or by the users of this product category and industry? What dependencies does this product have on internal teams, which may affect that delivery schedule? The answers to these questions determine the outcome the leader wants to have, which then guides them in selecting a new or revised methodology, understanding the methodology at the principle level, choosing the portions of the methodology that address the most significant delivery challenges and help achieve the desired outcome, and evolving that methodology adoption over time.

Similar to the previous three principles, which all have

distinct levels to their maturity models that can be reached and advanced, this principle is also divided into three areas of focus to be pushed on iteratively and advanced incrementally to increase the momentum of the team:

- *Understand product-specific work breakdown and delivery commitments.*

- *Define a feature-based squad structure that minimizes cross-team dependencies; establish the minimum leadership structure needed to guide the squads and disciplines.*

- *Improve team skills in the areas of estimation, prioritization, and sequencing.*

The principles we describe are hard pivots for many product leaders at large enterprises, and this mindset is best developed through a maturing process that is incremental and useful at each level, like any Agile methodology would deliver. For milestone purposes, we are prescribing three levels of maturity in developing this principle (and all others that follow), making it easier to imagine a continuum or evolution rather than often-tried and often-failed pivot.

Customer Delivery Expectations

To increase momentum in terms of the organizational structure, Agile methodologies must be tailored toward delivering to meet or exceed customer expectations.

In this focus area, product leadership teams will first benchmark current Agile adoption. Then they can seek to under stand customer expectations for the product they are

Evolution	Level One	Level Two	Level Three
Customer Delivery Expectations	Baseline understanding of product delivery patterns and customer feedback exists.	A process exists to make changes to Agile ceremonies to align better to customer needs.	A customized Agile framework to tailor to customer and product-specific needs is created and rolled out.
Feature-Based Squads	Product engineering teams are reorganized as self-sufficient teams.	Feature-based squads can release complete customer experiences without external help.	Truly elastic product teams can scale up or down based on the product lifecycle.
Feature Prioritization	Product engineering leadership has a formal repository of feature estimations and steps of doneness.	Product engineering leadership can manage multiple large teams through scaled Agile ceremonies.	A forum for constant sequencing and prioritization of features exists and is integrated into Agile ceremonies.

building and managing so they can determine the work breakdown and the right cycle of delivery to meet commitments. For example, the user of a retail product may have the expectation of delivering features to a customer continually, such as regular updates being released to an application. On the other hand, users of an institutional B2B product may understand that regulatory compliance requirements could affect the pace of new feature delivery and may not share the same expectations that retail customers do. Understanding the expectations of the user and the environment in which a product is operating can help calibrate the product engineering team's delivery processes to align with those expectations.

One of our clients was contracted to develop a new

WATCHING THE PLATEAU

At times, a leader may find that a team's answers to these probing questions don't necessarily highlight a specific hotspot. The team arrives at a position that all is okay, and there are no real issues to speak of. In these moments of plateau, it is important for a leader to ensure complacency doesn't creep through. High-performing teams need aspirational goals to strive for, to end the year better than where they started as a team. A leader must push and probe to help the team where they can stretch and grow. At certain times, this manifests in the creation of platform capabilities that can help multiple peer product squads. Other times, this can manifest as improvements in their Agile adoption and execution, and these best practices can be critical to other teams. It is only by stretching that people grow and are energized. Leaders may maintain the balance of this stretch zone between not letting them dip below to underperforming, nor allowing them to plateau, and not overcommitting the team, leading to missed expectations.

product for one of their customers, and they were contractually obligated to deliver the software to their customer by a specific date. With only a surface-level understanding of Agile, it would be highly unlikely that they could deliver to that date because Agile delivery frameworks don't generally ascribe toward hard dependencies on date to deliver. Instead, these clients had to make changes to the way they operate and plan out their delivery approach over the next nine to twelve months—a generally anti-Agile concept, but one which was necessary to meet customer expectations. They can still be Agile, but they couldn't be Agile for Agile's sake; they had to work backward from the customer expectations, choosing the Agile aspects that made sense for them but also incorporating other elements that allowed for success in delivering to a fixed scope.

Retrospective questions enable product leaders to gauge the health of their current Agile implementation within their product team and provide a sense of whether the team is in a more advanced state of execution:

- *What was the product roadmap for the last six months?*
- *How did the team deliver against that roadmap?*
- *What is the roadmap for the next six to twelve months?*
- *How is the team tracking toward that delivery?*

Additionally, in an enterprise construct, product leaders can solicit feedback from their counterparts taking the products to market so, if those counterparts experience pain or shortfalls in that delivery, they will make those shortcomings known. The product leaders can consider that an indication that further analysis is needed for their portfolio as well.

Follow-up questions should include:

- *Why did the team not deliver against the previous roadmap? Or why is the team not on track to deliver against the current roadmap?*
- *What methodology is being used?*
- *What is the rationale for using that methodology? If they are not using one at all—or the team is not completely clear as to which methodology they are using—why not?*

Each team should be familiar with some semblance of a roadmap or product release guide and should be able to articulate basic delivery constructs.

The product leadership team may also ask about other key metrics, in terms of backlog, having an understanding of team's capacity and therefore the ability to assign them work, and measuring velocity—a function of how well they meet that work assigned to them. New product leaders will want to understand what methodology their teams are following—or not fully implementing one, as the case may be.

Those interviews may also provide opportunities for other basic data gathering, such as, "What opportunities do you see for this team?" or, "What are the problem areas?"

All of this gathered information allows the product leader to triangulate real delivery issues a team is experiencing.

- *Does the methodology and roadmap approach align with the customer and internal partner delivery expectations of that team?*
- *Is the implementation of the Agile methodology correct?*
- *If the methodology and its implementation are acceptable, and the team is educated on that Agile methodology, but they are disordered in how they follow it, a leadership problem may be at hand.*

As an example, weak leadership results in common product delivery challenges that seem attributable to their scaled Agile framework. These can include the leader's inability to manage expectations: a leader who has set expectations unrealistically high is setting their team up for failure, disappointment, negative feelings, or taking on too much work and thrashing the team with expectations that can't be met.

A Bain "Agile at Scale" research study published in *Harvard Business Review*[20] concludes a concept called "taxonomy of teams" that maps customer needs to business processes and then to technology support while building roadmaps, and our principles are aligned to follow this approach, from business context, business architecture, and enterprise architecture.

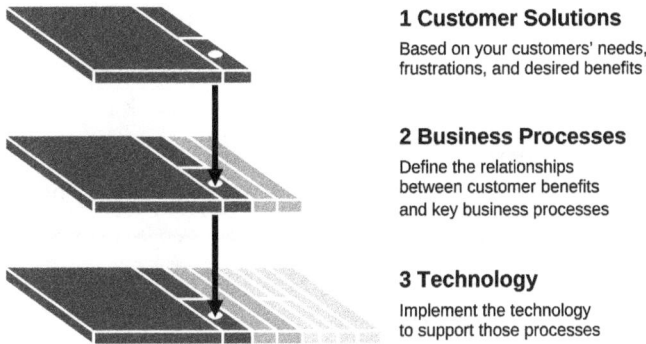

1 Customer Solutions
Based on your customers' needs, frustrations, and desired benefits

2 Business Processes
Define the relationships between customer benefits and key business processes

3 Technology
Implement the technology to support those processes

Figure 17: Mapping of Agile teams to business value streams

The same study reinforces our belief that deploying Agile the wrong way has dramatic consequences for an enterprise. Our example at the beginning of the principle developed a wrong "taxonomy of teams"—they are misconfigured to deliver customer value.

There may also be situations in which both the methodology and leadership are effective, but the physical constraints of the technology only allow for a ceiling level of velocity. (The product leadership team may have gained this insight from performing enterprise architecture and system architecture assessments in Principles 1 and 2.) In that case, the product leadership team may have to make the case for investment in removing that technical debt to unlock philosophy and potential from this team.

In this dimension, successful product leaders will have the information necessary to summarize and synthesize which teams seem to be doing well and which are not, particularly those that are impacting a major capability on the roadmap in the near term. It is unlikely that the product leadership team will get the executive buy-in necessary to bring change to the whole organization to fix some of these issues wholesale; instead, they will most likely start with key capabilities and then progress across the organization.

Every product has its own unique work breakdown and customer delivery patterns. Evolving the team's skill at developing a product starts by understanding these patterns, then making changes to existing Agile ceremonies in order to better provide feedback from customers. The final step is customizing Agile framework practices so that they are tailored toward each individual customer need.

Feature-Based Squads

The second dimension of this principle is determining the best way to structure teams to reduce dependency and cross-squad chatter within a sprint of work.

What is a feature-based squad? This is a self-sufficient team delivering customer-facing functionality. But what is self-sufficiency? This is not to mean squads will never need to interact with other teams, only that they will not be dependent on peer or vertical teams to deliver their features. It is common that shared services teams or platform teams help to provide subject matter experts in their areas, for example, cloud security teams or big data platform teams.

Features can be released by a squad independently. More

complex features might require multiple squads to contribute. Self-sufficiency means these squads can develop their portion of a feature and move on to subsequent items in their backlog without the need to hold. Coordinating release readiness of a feature once all contributing squads complete their portion of the work generally falls on a central release management role (in the case of SAFE, this would be the designated Release Train Engineer (RTE)).

Certain constraints may make it difficult for teams to achieve the highest level of independence—perhaps the marketplace does not support hiring a user experience designer for each squad, so one has to be shared across squads, thus introducing a dependency—but those constraints may simply be a reality of how the organization currently functions. This dimension of the maturity focuses on product leadership teams pushing to achieve a feature-based squad structure with minimal cross-team dependencies.

We worked with an enterprise cloud enablement team responsible for creating the cloud infrastructure the enterprise needed to support all their different application teams. The problem arose because their team structure was organized around specialized parts of cloud enablement; for application teams to get services approved from a cloud provider that they wanted to use required multiple hops. Internally within this enablement team, there was also a lot of cross communication to hand off a particular request from one of their internal customers.

Once this team went through an evolution of reorganizing their teams by product—in this case, by cloud provider—that allowed an end-to-end delivery cycle and an end-to-end view for their customers to see when they would receive their feature requests this internal team approved, built, and delivered. They saw an order of magnitude reduction in cycle turnaround time.

DETERMINING SQUAD SIZE

When going team by team, the size of a team supporting the product should be evaluated. More important than the size of the overall team is the size of the independent squads because a sweet spot to team size yields maximum productivity. As connections multiply, the number of people who need to be kept up to date also increases, which lowers the effectiveness of a team. Imagine the daily standup for a twenty-two-person squad; that team will find it painful. Large teams taking on many tasks can lead to confusion about what is expected and delivered, so it will take longer to assess the root cause of any issues. A product leader will be able to spot inefficiencies faster with teams performing smaller tasks.

Scrum teams typically have a best-practice range of six to ten team members, including the scrum master. Beyond that, it's difficult to determine what everybody is working on. Amazon has the concept of a "pizza team"—that the team should be no larger than the number of people that two pizzas would comfortably feed.

These are not hard and fast rules, but a litmus test to be cognizant of as teams are evaluated. This presents an opportunity to potentially reorganize a team if it is found to be constrained by its own size.

Feature Prioritization

Once the product leadership team knows at what cadence they need to be delivering their product and how their teams are structured, this dimension focuses on how to help those newly formed teams own their work by improving their ability to estimate, prioritize, and sequence work correctly. With continuous improvement in these areas—with a positive feedback loop of re-estimating, reprioritizing, and re-sequencing—product leaders will see results in terms of the accuracy with which they can make a promise to the customer and then deliver against that promise, having the right impact on the business by prioritizing the right work, and sequencing their work correctly to avoid collisions or wait times between squads and teams, especially where one body of work feeds into another.

We worked with a product team building a fixed-income trading application where we noticed a red flag in the original incarnation of the product roadmap: their roadmap was not backed by epics and stories that had been accurately estimated by the team, rooted in numerically backed data; it was more of a directional roadmap with opinion-based estimates from very few people on the team, mostly the leadership. We were able to help them first break down the work into a more comprehensive list of epics, features, and stories, and then we taught them how to do a true bottom-up estimate of the work using the knowledge they had based on different dependent parts of the product, past experience, etc. They could then use that data to build up an accurate total effort for each feature and sequence those features in a logical plan that gave the team the most efficient method of operating while also enabling teams to work on features with the least amount of downtime.

The net result was a far more accurate delivery plan with, unfortunately, a much later delivery date—the original plan had been an order of magnitude off base at critical milestones they had to meet with a vendor—but a more defensible date because the team could back up any questions about the extended timetable with quantified numbers that were now based in experience, not opinion. It also equipped the product owners and leadership to make trade-off decisions on scope because they could see the exact magnitude of effort it would take for one scope item versus the other and re-sequence or reprioritize as they saw fit.

ESTIMATION

A common issue we have seen is product leaders making inaccurate estimates, either too aggressive or overly conservative, which leads to missing targeted deadlines. The more confusion or opacity that exists around a plan, the more people will try to force as much work as they can into it; however, with increased visibility and transparency, trust is built and more reasonable expectations can be established.

There are several methodologies available specifically for estimation, but a combination that we have seen to be effective is a planning poker exercise and program evaluation review technique, or PERT, analysis.[21]

A planning poker exercise is a rudimentary way of conducting larger-scale estimates relatively quickly, based on the wide-band modified Delphi method. In this exercise, an entire team looks at a body of work, asks any questions they may have about the work, and then they each give their own estimates on a day's level. Outliers are discussed as a team, with the ultimate goal of triangulating on a final story point estimate for that piece of work. This exercise is typically performed for all work that a team or squad takes on, with every squad performing the exercise, so the leader has their first aggregate estimate of the entire body of work. The upside of this exercise is that it typically involves the whole squad and leads to quite nuanced estimates based on the epics. The downside, especially at the onset, is that the team may have a substantial volume of work to perform, so attempting this planning poker exercise for a large backlog will take an extensive amount of time. In that case, the leadership team would want to focus on the second methodology.

The second methodology, a PERT analysis, is more template-driven, wherein the team evaluates what kind of component they are building and the steps of doneness necessary in order to build that particular component, and then they graph a critical path toward completing those steps of doneness—perhaps their work takes four days, but they have to work with different groups, so they allot ten days overall. Because this exercise is more templated, teams do not need to debate time across thousands of stories, epics, and features. Instead, the squad considers the type of work to be done and applies the appropriate template for estimation. PERT is an industry standard for large program estimation; a technique based on PERT called three-point estimation[22] could be applied to programs spanning multiple quarters and teams, so it can be used to plan effectively for those organizations.

SEQUENCING

Once teams have estimated units and blocks of work, the product leaders need to determine how their teams will deliver that work. A process called Program Increment Planning, or PI Planning, part of the SAFE methodology, is one example of a centrally facilitated exercise for a product team to build a plan (though many variations can be used).

With this methodology, the product leadership team lays out the work across the delivery plan and begins to draw out the dependencies so they can then adjust sequences. It may not be possible for a squad to simply deliver work in a completely uninterrupted sequence because, once that work is viewed in its entirety on a shared visual, it becomes clearer how those dependencies affect the overall plan for the work. Once the leader completes that first pass, they can then try to determine why so many dependencies exist—is it a team organization issue, where a significant portion of the skillset is concentrated in one team, leading to multiple dependencies on that team? Is it a work ownership issue, where it may make sense for work to shift amongst squads so they can own it end-to-end without a dependency in the middle of the flow? That product leader may see a range of ways to alleviate some or all of these dependencies.

PRIORITIZATION

Prioritization occurs, from a planning perspective, (1) when people are bought into the total effort necessary to complete the work, and (2) when people understand the finite limits of the dependencies they have. Then, the most realistic plan to delivery can be created.

Depending on the circumstances, different teams may

SEQUENCING LARGE PRODUCT FEATURES AT SCALE

Large product rollout requires coordination with multiple independent squads all working toward a common mission. Once each of the squads estimated work on their own, teams would then gather in a large planning room to extend this exercise: a giant plot is posted on the wall with all of the sprints for that quarter, and all the squads use Post-it Notes to represent their sequences of delivery. Then, threads are used to show dependencies across squads. After all dependencies have been determined, the leads examine the quarterly delivery plan to determine the work ahead with a focus on how to remove as many of the dependency threads as possible, because each thread represents an impact to the flow of momentum. While it is unlikely they will be able to remove every thread, the visual nature of this exercise within the PI Planning process helps people to see and understand the impact of dependencies, become more comfortable moving work around, and removes some of the territorialism that may be present.

During the global pandemic, delivery teams around the world moved quickly to adopt digital planning tools to mimic the in-room planning experience as close as possible.

choose to prioritize work differently. In one context, they may choose to prioritize work in such a way to minimize as many dependencies as possible; in another context, the business may choose to prioritize getting more economic value out of the team and into the customer's hands as soon as possible. In larger or more complex systems where nothing is taken to the customer midway—for example, there is likely no midway release of software for a satellite—sequencing might focus more on where there is more technical risk, front loading that work so they don't bite the team later in the delivery cycle.

Agile begins and ends with teams, so a common mistake to avoid is attempting to apply an enterprise-level Agile view down to those teams. The foundational principles of Agile are to empower teams to do whatever they need to do to meet the customer needs as fast as they can. Beginning with that philosophy of empowerment means also beginning with a team-level discussion and assessment. Once necessary optimizations have been identified at the team level, then the product leaders can begin to roll up to determine what additional frameworks the teams may need or what else may be needed to fulfill other standard, inescapable enterprise bureaucratic needs.

Achieving success in all three Zones means achieving the ideal state discussed earlier in this principle—teams are happy and feel good about making and meeting commitments, feedback is promoted and shared, everything is clicking, and teams dominate the marketplace because they're able to move so quickly. It may seem difficult to imagine, but this is achievable; success is a function of time and investment, similar to a karate student taking the time to master the steps necessary to earn a black belt. It also takes time for teams to learn about each other, gel together, and find their rhythm. With the right people, the right leadership, and the right framework, eventually they will get to a point of being highly predictable in terms of what they can deliver.

COMMON CHALLENGES IN IMPLEMENTING THIS PRINCIPLE

Implementing new frameworks without considering full organization changes.

Engineering organizations building software products have been using methodologies to deliver software for a long time, and, historically, teams at these large organizations were organized into silos. When introducing a new feature, they would extensively plan for the ultimate delivery date to the customer. This effort was effective until organizations began to implement Agile universally, at which point enterprise organizations typically brought in new frameworks and may or may not have also changed their existing organizational structure. Lack of medium-term (quarterly) planning and product roadmap strategies, due to applying textbook Agile, led to significant thrashing between teams as everybody was moving quickly but nobody understood how they were supposed to work together to deliver software within the context of how the organization works and how their customer expected those releases.

Lack of depth of understanding of framework nuances.

Many leaders have a "check the box" mentality: they understand the Agile framework from a theoretical standpoint and so are able to honestly say they implemented a framework with their team—instead of examining how best to apply the framework most effectively for that team. These Agile frameworks can only work if leaders understand both the framework *and* the expectations of the organization's customers. Then they can choose the best framework and change the organizational structure for greater efficiency and increased momentum.

Resistance to structural changes.

Without a doubt, the main problem a product leader is likely to encounter as they implement this principle is resistance to structural changes—and that aspect can make this principle particularly challenging. In previous principles, issues can be distilled to simple areas such as capabilities, systems, or data to be changed or fixed, none of which have emotions or egos to take into consideration. Unlike deciding whether or not to retire a mainframe, here product leaders may have to make difficult decisions that impact *people*, not technology. Particularly in an enterprise setting, people have often worked for the organization for many years, even decades. A common byproduct is that teams have broken down their work by domain instead of by end-to-end customer capabilities, which is more in line with Agile. Newer frameworks also push toward more integrated teams, which requires changing ownership and the makeup of those teams, and which can lead to pushback as team members tend to want to preserve the culture and unity of that team. Leaders may have to make difficult decisions about the best way to organize their teams such that customer features are prioritized, not defaulting to how the organization was already structured.

Inconsistent adherence to the firm methodology across squads and team members.

We performed an assessment for a client with one squad where the feedback indicated low adherence to methodology. After a discussion with the scrum master, we determined that one member of her nine-person scrum team had a thirty-year tenure with the firm and held vast amounts of knowledge about the back-end systems. The scrum master felt that she had limited recourse if the legacy team member chose not to attend the daily standups because the organization would not want to upset someone with that much institutional knowledge. Thus, she allowed him to skip the daily standups, which then signaled to the rest of the team that those standups were not important, so many of them attended inconsistently, if at all. Key-person risk needs to be addressed quickly because it can easily spiral out of control, as was shown in this instance. When a team's leadership is not as disciplined in maintaining rigor on certain processes, the whole organization suffers.

Managing expectations with team members on the organizational evolution that will occur.

An additional challenge a product leadership team may face when implementing this principle occurs in setting *realistic* expectations for what can be achieved with the Agile framework. Junior members of the team may attend Agile instruction where what's taught is the most utopian implementation of these frameworks—teams have total autonomy, they're multidisciplinary, they can deliver features without consulting other groups, and they have complete control over their schedules. When those team members return to the enterprise and expect to implement Agile exactly as they were taught, they are dejected to learn that reality does not match up to what they learned in class. Part of the product leadership team's role is to calibrate between the ideal implementation of the framework versus where their organization is today and how far they can take that organization in the near term, with an eye toward where to take their team and sibling teams in the organization over the longer term.

Not accounting for iterative review and course correction of the implementation.

Another common challenge comes when an organization decides to embark on an agile transformation and product leaders have the mindset that a framework will be applied once and then need no further change. Even if the chosen framework was customized and the leader performs their best faith effort to implement it with the right orientation for that organization, there is no way of measuring whether the new teams are delivering what was anticipated and, if not, where the problems exist and how they should be addressed. Instead, the correct way to approach Agile implementation is to have a feedback loop to iterate on the framework until optimal results are achieved.

Sustaining the change transformation effort.

Finally, it can take years to build the trust that teams have with their senior leaders, to get to the level of autonomy preached in many scaled Agile frameworks, where teams can decide when features go out and have delivery dates communicated bottom-up instead of from the top-down. Lack of executive buy-in, or a lack of interest in unpacking frameworks and working through existing issues, is a challenge product leaders are likely to experience. When this occurs, it often

LACK OF DIRECT CONTROL

Outside of the core challenges product leaders can control, there are a few other challenges that are outside of product leaders' direct control:

First, the budgeting process is different in an enterprise context, which scales up teams as necessary and releases those additional squads once the work is completed, as opposed to a product firm, which funds permanent product teams in perpetuity. That fundamental difference in a product mindset means that, rather than focusing solely on measuring spending, now the measurement shifts to ensuring that the efficiency of the team is increasing over time and that the product is meeting whatever business outcomes are desired.

Second, dependent teams are another consideration. Whatever one leader does with their team may not be what all the teams are doing in the context of an enterprise, so how can product leaders manage the dependencies and drive that through influence and trust as well? They should realize that they are likely not going to be able to effect change among those other teams, which may be more or less Agile than their own. While they cannot change them, they can influence those teams because they are reliant upon one another to deliver the end product. Teams are trying to achieve the same customer-related outcome, so they should work to be in alignment in terms of deliverables, with leaders being aware of the touchpoints between products and planning for respective handoffs or incoming work.

results in issues getting buried, which can only happen for so long before the consequences are felt.

The most basic level of framework implementation is the path of least resistance for executives. Business owners do not have a deep understanding of software delivery, but product leaders need their buy-in to effectively implement tailored Agile frameworks, so those leaders will need to introspect on the benefits of the new framework—such as better team morale and improved customer experience—and be able to present them to the business. Product leaders who embrace this principle should also understand and calibrate to the level of buy-in their executive team has on these frameworks so they can manage expectations with their teams on the realities of their business and how far they can move forward right now versus other tiers of maturity that can be approached over time.

CASE STUDY: LARGE FINANCIAL INSTITUTION

A large financial institution, with six squads of approximately ten members each, needed to deliver a large payment platform. The team endeavored to use Agile practices in dynamically allowing squads to select and prioritize their own feature delivery.

The team ran into a challenge with external fixed dates that had to be met to avoid legal and financial liabilities. Initially, the team was concerned that the external construct/dependency would force them into following a waterfall approach to assuage executive concern on the specificity of delivery dates.

In working with the team, we were able to define a hybrid

approach that took advantage of the sprint cadences, continuous customer demos and feedback with added planning and mitigation to manage scope and delivery risk. Reprioritization occurred with the customer to ensure features required by the deadline were delivered.

If a textbook Agile framework had been followed, the executive team would have been in the dark on the percentage completion toward the minimum viable product (MVP) needed to avoid the legal milestones. Instead, the entire backlog was estimated bottoms up. That drove the number of squads, prioritization, and the plan.

By nurturing the organization to develop truly fit-for-purpose product engineering methodologies, we have noticed that organizations get excited to tackle transformative programs, create quick wins and trust, and build an actionable roadmap. Next, we will look at how product engineering leaders can efficiently channel this energy to build and roll out resilient products.

KEY TAKEAWAYS

1. The value of long-term retrospectives of enterprise Agile and customer feedback cannot be underestimated. They are an important foundation to help plan for future needs.

2. Customer needs should be the basis of team organization, execution, and delivery processes.

3. Prioritizing features is doing the right things and ensuring they have a clear line of sight into both expected customer and business impact.

4. When customizing any Agile framework, one must factor in organizational size, scope, and architectural constraints. No framework works off the shelf; it has to be customized for an organization's particular needs.

5. To scale as an organization, product leaders have to strive toward creating self-sufficient Feature-Based Squads supported by platforms and infrastructure teams.

PRINCIPLE 5

OPERATIONS AND TECHNOLOGY

Invest as Much Time in the Health of the Journey that Produces Your Products as You Do in the Product Itself.

"The issue is not about coming up with a car design—
it's absolutely about the production system. You want to
have a good product to build, but that's basically
the easy part. The factory is the hard part."

—ELON MUSK

When Elon Musk built the Tesla factory, he devoted as much time to designing the factory as he did toward the design of the car because. if his cars were successful and in demand, then speed from order through delivery would be an important factor. The factory is designed to not only get cars out as quickly and reliably as possible, but also to make sure they are free from any defects. The plant has been specifically engineered so that the process of assembling vehicles goes smoothly with no bumps in the road for workers or buyers alike.

For the product leadership team at an enterprise organization, the car in our example represents a new business capability, and this principle around Operations and Technology is focused on the factory where those products are built and the environment in which they operate. Coming back to our car example, once the car is delivered, product teams want to monitor the usage and health of the vehicle[23] to provide continuous improvements and safety functions. This capability in our software case is observability of the product and post-production change management.

A product leader is continually allocating budget, so they have to decide for their portfolio whether it's most important in that moment to improve the factory or add new features to their product. Product leaders have historically had a difficult time making the case to investment committees to invest in these foundational platforms, but due to the digitalization of industries and highly visible consumer experiences, it is becoming increasingly necessary to fight for these budgets.

THE IMPORTANCE OF OPERATIONS AND TECHNOLOGY: BUILD AND OBSERVE

This principle focuses on how to construct the right factory to build the product and the techniques to use to operationalize products for continuous monitoring, customer feedback, and resiliency. Without this fundamental pillar, product organizations will be flying blind to the interactions with the customers and real world.

This principle in a broad sense is typically shared between Technology (production monitoring metrics), Sales and Marketing (user and visitor tracking metrics), Product (feature metrics, conversion funnels), and DevOps (build and deploy metrics). The success of the product depends on product leadership having an overarching view of the end-to-end factory that makes and runs these high-quality products.

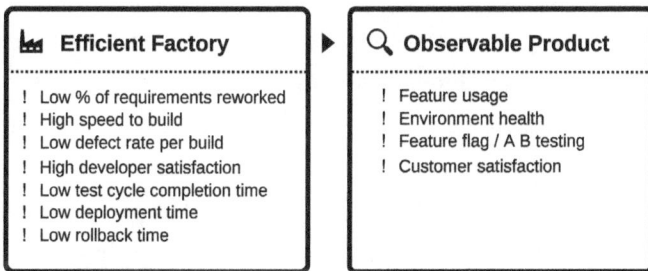

🏭 Efficient Factory	▶	🔍 Observable Product
! Low % of requirements reworked ! High speed to build ! Low defect rate per build ! High developer satisfaction ! Low test cycle completion time ! Low deployment time ! Low rollback time		! Feature usage ! Environment health ! Feature flag / A B testing ! Customer satisfaction

Figure 18: Dual focus on achieving an efficient development flow, looped back with usage data from production, enables a positive improvement feedback loop

It may be helpful to think about this principle as two sides of the same coin: on one side, innovation via technology—

meaning high-quality factories producing high-quality products; on the other, instrumented products to make the product observable once in the hands of the customer. This is a horizontal capability that touches the entire lifecycle of the product from design to production.

In large enterprises, there has been a shift in focus in measuring the efficiency of the software development lifecycle (SDLC) and technology operations. Product leadership should have a baseline understanding of how business capabilities make their way through the various stages (Development, Testing, Build, and Release), what service levels are guaranteed when in production, and which metrics are important to monitor and track improvements.

In the previous four principles, we have put forth an examination of other areas of focus for a product leadership team: business context, top-down enterprise architecture, data, and people organized within an Agile methodology—but we have yet to discuss the software development lifecycle and efficacy of operations that are intermingled here in Operations and Technology. Product leaders should use the first four principles to determine strategic areas important to the business and technology teams, and will now use these operational metrics to get a sense of the truth behind product performance. These metrics ultimately are also a proxy to developer velocity, product quality, and the speed at which the business features are developed and deployed in an optimal fashion.

There are real bottom-line implications to investments in this area. In a survey of 440 large organizations, McKinsey research showed that those in the top quartile of developer velocity correlate with revenue growth that is four times greater than that of the bottom-quartile firms.[24] Top-quartile companies also demonstrated 60 percent higher total shareholder returns and 20 percent higher operating margins. Consequently, top-quartile players appear to be more innovative, scoring 55 percent higher on innovation than bottom-quartile companies. These businesses also score higher on customer satisfaction, brand perception, and talent management.

Build

The product and software development lifecycle represents the Build portion of the principle. A leader of a portfolio of products needs to have a pulse on the technology value stream, i.e., the factory, because it denotes the nimbleness of the organization—or the lack thereof—in being able to get products out to market. Our clients often experience gaps in this phase where the process is stalled or fragmented. They have the right capabilities, good architecture, and the right team, but perhaps due to a lack of maturity in the product build process, they are not able to develop and deliver products quickly and efficiently. This affects their ability to be responsive to the market—if it takes three months to get a new idea to production, that lag time will allow faster-moving competitors the opportunity to advance.

A product leader relies on a high-level understanding of the software development lifecycle—the process necessary to go from having an idea on paper, to building the new capability, through to implementing and making it available for customers—that a technology team follows. While the product leadership team doesn't need to understand *how* to create the capability, they do need to understand why it is important, what it takes to build it, how their teams, people, and technology are organized around the idea on paper, and the expectations around deadlines so that they can understand the metrics that might be presented to them and make the right commitments to the business.

A hypothetical example may be deciding to add a new feature in an application that, when used, shows customers their account balance history. When estimating new features, often there exists a dichotomy of its implementation complexity. On the one hand, instinctive estimation could suppose that the feature should be quick to implement. When exploring the development lifecycle, however, understanding the time to develop the feature, add it to the product web and mobile applications, storing large volumes of account balance histories, testing effort, and production A/B testing time may yield a different estimate. Reconciling the variance in both these expectations could uncover areas of improvement to the overall build process.

The above is a simple example, but consider a more extended scenario (as it is likely to be for large organizations): let's assume that the product leader has a product available in the US and is now tasked with launching that same product in Europe. In this scenario, the product leader would need to consult with their technology team to understand the breadth of changes needed to support this requirement. A complex undertaking like that type of geographic expansion will likely have legal, operational, and technical considerations. The need to isolate payment and privacy data could imply a need for a local data center. Product user experience changes may also need to be introduced to support the requirement.

Whether adding a new capability, like the previous

example, or changing how a particular capability is being used by a customer, from the standpoint of a product leadership team, the five steps of build doneness to consider surrounding the software development lifecycle are:

1. Have we defined detailed enough requirements on what we are building? (We have seen cases where the teams are just iterating through requirements, which will delay the release time.)

2. What are the steps, within the software development lifecycle, in terms of building this capability? (Develop, Build, Test, Iterate, Stabilize, and Deploy)

3. What are the steps for testing the capability?

4. What are the steps for accepting that capability?

5. What does it take for us to move this into production so that the customer can use it?

One often overlooked element of value stream is strict scope management. Large enterprises have a culture of building a final product to be launched versus more nimble organizations that build an MVP and iterate after public release. What constitutes an MVP is not just a product, business, and audience-specific question but also a cultural one. Amazon is notorious for releasing functionality in infancy and aggressively iterating based on real-life customer feedback. Not every industry or vertical would have such a luxury. For example, a financial services application dealing in high dollar transaction amounts cannot afford to iterate heavily on their core product, so those capabilities may need to be whole for the MVP. Other parts of the product, however, like ancillary reports and features could be progressively rolled out past MVP.

Overall, the answers to these questions about the build value stream determine the amount of work to be done. With that established, product leaders can also inquire as to the velocity at which this feature will be completed, allowing them to estimate the amount of time required to deliver the product. If the tech team says that a new feature is 1,000 story points (typically a story point is assigned to be a one-person-day-equivalent effort, but it is a convention) and they have ten people working on it, they know it will take one hundred days to build that capability. The product leader can confidently tell their management, "We'll have this feature ready in six months." With that understanding, the product leader can monitor how many story points have been completed and whether the team is moving at the same velocity throughout.

At this point, the product leader should have a roadmap showing all the changes in progress and in which upcoming quarter those changes can be expected to be delivered. From there, the product leadership team will examine their software development lifecycle change-management view. For each team working on new developments, the product leader will want to know how they are trending on a weekly change management dashboard, which should be an "on the wall" artifact and could be delivered electronically. This dashboard shows how far the teams have moved toward the goal of that six-month or one-year target, typically indicating a project status as green, amber, or red.

Any projects that are not green on the change management dashboard indicate potential risks or issues that have arisen in terms of the release of that feature—and they require further investigation as to the root cause of the bottleneck. If instead of taking six months, as was originally determined, a project is now slated to take eight, at the top level the product leader knows simply that the project is not progressing according to plan. Here, they can determine what is causing the delay—is it because they weren't able to properly document exactly what was needed? Is the technology team not finishing program increments at the necessary velocity? Is the team finding unanticipated defects and issues as part of the continuous integration process? Or perhaps the development team is delivering as promised, the requirements team indicates that all the functions are working, but the pilot customers or business users are coming back with unexpected changes.

The product leader will address each of those four different scenarios in a different way. For example, the first scenario is due to a requirements mismatch, so the product leader may then ask their technology lead:

- *"Can we make something simpler?"*
- *"Can we take something away, in order to make the timing work and hit the deadline for going live?"*
- *"Can we have some of our customers do testing early so we can find these issues sooner rather than later?"*
- *"Do we need to add more people to an existing squad? Or do we need to add an additional squad?"*
- *"Do we need to increase test automation coverage? Should we increase automated regression testing to decrease time lag to deployments?"*

As product leaders move deeper into these areas of concern, they are ultimately responsible for getting the product to market, and they must make decisions in terms of their commitment to customers and market expectations of the business, so they need to work closely with their technology team to remediate any issues.

By continually fixing the bottlenecks that will initially arise, the faster each product team will become, ultimately leading to the end state of an incredibly strong team that is highly responsive to the market. At their peak performing state, as we discussed under our Agile organization principle, a product team can ingest requirements, break responsibilities into manageable chunks, prioritize those tasks in a way that has the highest impact on the business, get it quickly through the design to delivery system, test it appropriately, and put it in front of the customer in a safe and reliable manner—and they can do it all quickly and with minimal, if any, friction. Every product team should strive to embody that vision.

That vision keeps product teams on track, yet most often, they neglect to return to this vision. Many product leaders gravitate toward the important, but separate, topics of what features should come next, faults with the design, or firefights with other areas of product. They rarely step back and evaluate whether their development process is as efficient and effective as possible. Continuous improvement is key to ongoing success within the build phase.

Observe

Production data is the source of truth. Unlike any hypotheses someone may have prior to a product going live, data collected from production is free of bias; it is authentic data. There is considerable value in data about product usage—customer's point of view, who is using it, how often they use it, where they use it from, how they flow inside the product—in terms of investing in and improving the product. First, that usage observability of what is happening to the product needs to be captured so the engineering team can improve the product.

Beyond this data capture, product analysts need to look beneath the surface of the business metrics, such as conversion rates, customer order funnels, etc. Technology and Operations teams need to understand how each service is behaving and the best way to recover due to inevitable failures. Some of these advances are termed as observability, the practice of which has been exponentially increasing since 2016, when the term was officially borrowed from the control systems world.[25]

Large enterprises typically never revisit a built feature, unless there are defects, and do very little in terms of feature tracking and customer dynamics on the site with introduction of new features unless there are real business impacts. There is a growing sense of urgency to develop real-time insights on product performance in the market, a practice that is common for upstarts and technology companies. Product leaders should constantly monitor search engine optimization (SEO) metrics, KPIs, A/B testing impacts, feature performance, and conversion rates using application analytics and product monitoring tools.

In addition to production usage, operations teams need to continually evaluate how the product is performing from an operations point of view—as an example, the availability of services, throughput and site slowdown, etc. But in order to do so, they need to have proper instrumentation to capture the data, and even most technically advanced companies have challenges in maturing to a level they can isolate major issues. For example, Robinhood, the investment and trading platform, was able to build an entire brokerage organization in less than two years; they are considered to be a well-executed machine, advanced in technology, and nimble in terms of releasing features. They have invested considerable resources on the backend to ensure trading spikes could be handled and the platform would remain stable. However, when their site went down for several hours in March 2020, it was due in part to neglecting to monitor the customers logging in and the spikes associated with that.[26] Robinhood said on their blog, *"We've taken steps to address the root causes of the March 2020 outages, reduce the risk of future outages, and increase the resilience of relevant systems, including by increasing system redundancy, better distributing load on Robinhood's systems, and deploying a risk-based testing system."* This example only serves to emphasize the importance of capturing and monitoring the health of the product along with the usage metrics in terms of how the customer is interacting with that product. It also demonstrates that no one is shielded from failures, and it is how fast one can recover that matters. As AWS CTO Werner Vogels often says, "Everything fails, all the time."

We have earlier discussed monitoring the value stream steps within the software development lifecycle—the flow from idea to architecture, to concept, to build, to testing, and finally live; this affects market responsiveness. Operational issues such as product downtime and product responsiveness affect customer perception. With both of those pieces in the correct place, the organization can measure and improve the necessary metrics to move quickly and to be more responsive.

COMMON CHALLENGES IN IMPLEMENTING THIS PRINCIPLE

Common challenges in implementing this principle originate from the fact that the investments in improving software delivery (Continuous Delivery, Infrastructure Automation, etc.,) are not fully appreciated as business enhancing, hence the struggle to get investments focused in this area.

Peter Drucker once said, "If you can't measure it, you can't improve it." Thus, the implementation of observability is based on metrics that are relevant and move the needle on the health and usage of the product. But putting instrumentation and "sensors" on a product requires knowing the KPIs that matter, accounting for the engineering effort, and an understanding of the cost of downtime to size observability investments appropriately.

Common Challenges in the Build Phase

Lack of appreciation to develop a business case.

The first set of challenges relates to business buy-in. Continuing with the car factory analogy, improving the factory doesn't help a car salesman sell another car, but it has benefits down the road when car production is higher and quality is maintained, so it is equally important to provide a business case area where the factory should focus its improvements.

Legacy architecture impediments.

The product's architecture itself has to be designed in such a way so that the product features can be deployed quickly. At

Tesla, design options even for highly desirable features such as luxury interiors for the car get axed if they haven't yet figured out how to build those features at scale, automated, and with an acceptable error rate, whereas a lot of their competitors design interiors that may seem luxurious, but which are expensive to build and maintain.

Coming back to our product software architecture, large enterprises have monolithic architectures (mainframe, large set of features combined in a specific software component, black box systems that are old and no one wants to make changes, or changes take a long time testing) due to legacy. It is imperative for product leaders to recognize that to improve the factory, sometimes teams have to improve the non-functional portions of the product first. By thoughtfully retiring these monolithic patterns, product factory velocity could be improved and become nimbler.

Automation of product quality assurance.

The next challenge is the speed and automation of quality assurance (QA), which again occurs in the design of the product. Many modern technology product companies build their automation suites at the same time they are building the feature. It is not an afterthought; when considering a feature, they think not only about how to build, but how to build it in a way that it can be tested easily and in an automated way. This requires a high initial investment, so it is likely that this struggle will be improved incrementally rather than all at once.

Tools, tools, and more tools.

The last potential challenge in this area is being inundated by tools that help engineering teams be more efficient with software delivery. We have seen many clients start with the tool first, attempting to fit that tool into an optimum build and deployment process, instead of considering an ideal software build and deployment pipeline for their product and then finding the tools that help them achieve that.

One way to achieve this target state, from Principle 1, similar to business value streams, software value streams can be built to understand the bottlenecks within the factory. As Mik Jergerson, author of *Project to Product*, puts it, "Software value streams are not linear manufacturing processes but complex collaboration networks that need to be aligned to products. To avoid the pitfalls of local optimization, focus on the end-to-end value stream."

Prioritize tool selection and investment in areas that aid the health of the overall value stream. Imagine a full end-to-end process (say, automating the build and testing of a web application) is a flow, and each task within the build factory is a flow item.

Three variables can help measure and identify areas of tool investment:

- *Flow Distribution: The proportion of each flow item within a value stream, adjusted depending on the needs of each stream to maximize business value (build task, infrastructure deployment task, etc.).*

- *Flow Time: The duration that it takes for a flow item to go from being accepted for work into the value stream to completion, including both active and wait time.*

- *Flow Load: The number of flow items being actively worked on in a value stream, denoting the amount of WIP (work in progress).*

Properly calibrating factory-specific investments can yield meaningful impacts:

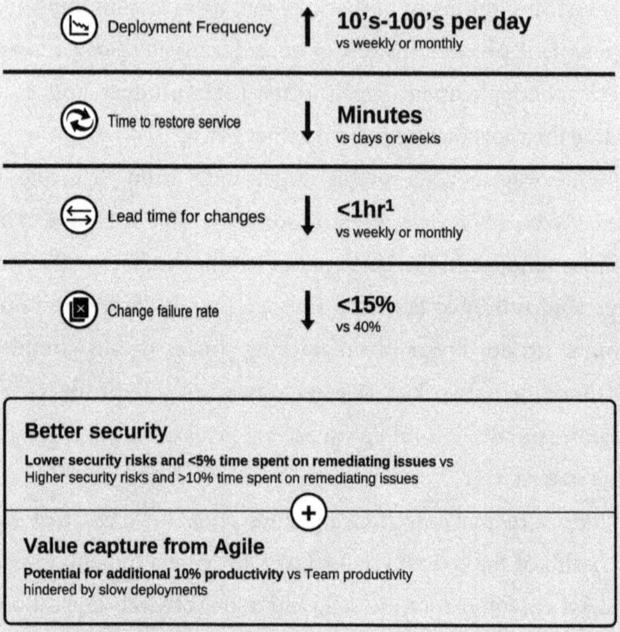

Deployment Frequency	↑	**10's-100's per day** vs weekly or monthly
Time to restore service	↓	**Minutes** vs days or weeks
Lead time for changes	↓	**<1hr[1]** vs weekly or monthly
Change failure rate	↓	**<15%** vs 40%

Better security

Lower security risks and <5% time spent on remediating issues vs Higher security risks and >10% time spent on remediating issues

(+)

Value capture from Agile

Potential for additional 10% productivity vs Team productivity hindered by slow deployments

Figure 19: Investments in factory health yield significant improvements in operating velocity[27]

Common Challenges in the Observe Phase

Underestimated engineering effort.

When building products, many product leaders underestimate how long it takes to incorporate instrumentation. This can hobble a team right at the onset if they aren't adequately budgeted or staffed to account for this workload.

Secondly, when trying to prioritize where to instrument

and measure, it can be a challenge to go through the mental exercise of considering every component and determining the business impact of that component not being used, being down, or performing poorly. This is a crucial activity to ensure the limited resources available for instrumentation are pointed toward the highest impact areas.

The other challenge is that engineering teams don't evaluate the cost of being unavailable. It can be hard for teams to pin down product service level agreements (SLAs) and component service level objectives (SLOs) within a product.[28] While this lack of requirements affects the product architecture, it also affects the instrumentation put in place in production to monitor the health of these components.

Building an engineering platform for observability is expensive and takes real engineering effort, which may not be accounted for in the core team delivery. Improving resiliency is an ongoing process, and best practices are around allocating a resiliency budget per squad, and demonstrating the need from actual production outages and manual interventions should be followed.

Business KPIs are not enough.

Product leaders should measure usage KPIs along with business KPIs. When features are rolled out, people may measure some of the business-impacting KPIs—such as how many trades are placed per hour or per day—or, when rolling out a new capability for cryptocurrency, tracking the obvious like how many Bitcoin people buy or sell and how often people check the Bitcoin price to gather user behavior—but the usage KPIs or user experience KPIs themselves are inconsistently measured.

Features that are not the main business transaction—such

as the path they take to place a trade, storing favorite tickers, etc.—tend to fall by the wayside, remaining in the code base with people supporting them, patching them, and rolling them out again, without ever evaluating if they need to keep this feature, whether anyone uses it or what they're using it for, and/or how they may be able to enhance it to bring about a different result.

One of our financial services clients who performs a lot of mortgage operations determined that their most-used feature was their mortgage calculators, but there was a noticeable fall out of traffic from this feature. The team redesigned the calculator and mortgage application integration experience and saw better conversion across the journey.

On the product usage side, the core challenge is in ingesting all of the usage data that can be collected. There are many instrumentation tools out there that are ready-made and allow for the creation of reports about how people are using the product and different web and API analytics, but without a framework to absorb that, product leaders can get inundated with a mountain of data.

HOW TO APPLY THIS PRINCIPLE

This principle can be applied to two areas within operations and technology: one area is the flow of the product (Build), and the second is the health of the platform (Observe). This section will first provide an overview of the maturity model in each area before going into greater detail about each individual level.

The principles we describe are hard pivots for many product leaders at large enterprises, and this mindset is best developed through a maturing process that is incremental and useful at each level, like any Agile methodology would deliver. For

IMMUTABLE PRODUCT— A CONCEPT BEHIND AUTOMATION

One of the modern engineering principles when it comes to building an efficient factory is building and maintaining an immutable product. What we mean by that is, changes should only be added or removed to the product, not edited, hence the term "immutable." It does not mean the product is not changing; it is changing in a less risky way.

This might seem counterintuitive to someone, but one of the biggest philosophical changes in software, with the power of automation, cloud, and virtual machines is engineering teams never have to "change" anything, as they are changing the product. Teams will remove or rebuild the server configuration and deploy the entire web application again, arguably to a completely new server or virtual machine. They could add and remove a few pieces of line in the software code and "retest" everything, including the new code that was just added with new source control mechanisms. It is no longer about editing configuration, editing software, or incremental testing, but it is about sending a completely new software to a completely new machine and discarding the old machine!

There are several practices that are aligned to this principle: Infrastructure as Code, Deployment as Code, DevOps, Continuous Build and Integration, etc.

milestone purposes, we are prescribing three levels of maturity in developing this principle (and all others that follow), making it easier to imagine a continuum or evolution rather than often-tried and often-failed pivot.

Because this principle contains information about Build and Observe, each level has been split into two subsections, one for each area of focus.

	Level One	Level Two	Level Three
Build: Software Development Factory	A Software Development Lifecycle Value Stream is formally created for the product portfolio.	The value stream is continuously optimized by leveraging DevOps tools.	Product velocity metrics are developed and monitored to understand the flow of change from requirements to delivery.
Observe Operations	Tooling and monitoring exist to observe the top-level health of the platform, including monitoring production issues and customer complaints.	Product leadership team has formally defined analytics and stability at feature level through service level objectives.	Product teams have formal resiliency guide-lines established, including business continuity and recovery. Teams can run experiments on features and obtain production user feedback through analytics.

Level One

BUILD

The first level of maturity is about understanding the software value stream of the product portfolio. The software development lifecycle is not a new field, so even the most nontechnical product leader understands the idea of requirements, definition, concepting, designing, building, testing, and releasing. However, even though they understand this conceptually, most product leaders fail to track the necessary metrics in a way they can then act upon to explore and remove bottlenecks—if they can't see these potential problem areas, the product leaders cannot adequately address them. Later in this principle, some of these health metrics are highlighted, including categorical and numerical metrics for build improvements.

Recently, we worked with a product team whose leader knew, anecdotally, that the development team would need three months to be able to release a new feature. The product leader had this cursory level of knowledge but hadn't instrumented the process to the necessary depth to ask, "Where exactly in the lifecycle are we slowed down? Is it a capacity issue or a development speed issue?"

While the question was relevant, there may be multiple root causes that are contributing to the issue. Additionally, causes of the bottlenecks will vary, so the mindset is less about diagnosing the specific issues, which will be different for each team or organization, and more about institutionalizing the aspiration of each project team to improve their flow. To do so, the product leader can measure the following in this level:

- *What is the rate of flow, right now, from end to end, concept to delivery?*
- *What goals does the product team have to improve that flow?*
- *Where do we believe the greatest bottleneck is?*

OBSERVE (HEALTH OF THE PLATFORM)

Operations is the source of truth about customer experience.

Traditionally, the only basic requirement for measuring the health of the platform is to measure the core components in real time: Are all platform components running? Are they healthy (CPU, memory utilization)? Have there been complaints about downtime? But the only time anyone looks at these is when there is an outage. Level-one maturity is about changing this mindset.

To have operational stability, a set of four major metrics must be measured and continually monitored by the product leadership team:

1. The top issues faced in the last month

2. Top issues faced in the last quarter, latest status, and appropriate remediations done during this month

3. Significant outages

4. Business KPI trends for customer value streams

Even for those product leaders building new products, incident management and operational readiness need to be top of mind—just not on day one. When building a brand-new product, historical data will not yet exist; as the product develops, however, the product leadership team will want to collect the data necessary to perform an analysis on it six months later.

Successful product leaders will invest in installing a dashboard to reflect a continuous stream of accurate, real-time data—allowing them to monitor the heartbeat of their portfolio.

Top Issues Faced in the Previous Month and Quarter

The first two metrics can be looked at, in combination, as providing the voice of the customer. Product leaders need to be aware of issues customers are experiencing, capturing that customer experience in a formal voice of the customer program, in order to better understand the quality of the platform and product. This input may be delivered haphazardly, from many different directions, but nonsystemic issues identified should be channeled into their queue.

The firsthand research provided by customers who are actually using a product or service is invaluable. Without this program in place, the organization will miss out on valuable insights from their customers on how to make their product better—including potential ideas that the product leader may never have considered.

The organization may already have something in place, but the voice of the customer should be formalized within the operations and technology team.

Significant Outages

With regard to the third metric, most enterprise organizations already have a large infrastructure footprint with control centers dedicated to continuous monitoring of their systems; however, we propose a shift away from reactivity and toward every product leader gaining the insight delivered from having a real-time dashboard of the business capabilities and service level objectives they support. There is a high likelihood that organizations have instrumented data to detect and escalate outages. Having that data is not a new concept; the differentiator here comes from measuring the live data with a view toward also analyzing historical data in order to anticipate outages in the future.

Even more challenging is the secondary monitoring, not of outages but of degradation that occurs slowly. Perhaps the time between trades lengthens between Monday and Tuesday, but not by enough to cause alarm. If that gap is shown to be significant when the data spans from January to June, however, it indicates that a degradation has been occurring over the previous six months. A product leader must determine the mechanisms that are in place to highlight whether a degradation in service even exists—*before* it becomes an issue.

Business KPI Trends for Customer Value Streams

The fourth set of metrics allows the product team to observe product usage. These metrics are specific to the organization, but this live data is an indicator of the condition of the capability—and even of the company within the industry. For example, retail trading application teams monitor "order abandonment" as one of their metrics; understanding why a customer chose not to move forward with an order may uncover issues with the user experience or the technical performance of that step. If we return to the Tesla example that opened this principle, Elon Musk may choose to monitor the defect rate for Tesla cars. Assessing patterns for why cars are brought back most frequently to the dealership may identify issues in the upstream manufacturing process or design that need to be addressed.

To determine where issues exist, product leaders should strive to develop a culture of monitoring and being responsive to business and service-level trends. As an example, one KPI may be the number of trades submitted in the last hour. A sudden 30 percent drop in daily trade volume could indicate a problem. Similarly, if the number of active users on the platform is below the normal historical, a problem may exist. Other business metrics may include the number of customers signing up or dropping off. Large enterprises such as Netflix and Amazon Prime showcase that these are public-facing company performance metrics.

Lack of a monitor and response culture is a common challenge, but product leaders are equally challenged to define *what* to measure, with most people measuring either too many metrics or too few. One client initially had nothing at all to observe but installed a new observability platform for the incoming product leader to monitor and then listed 850 metrics on their dashboard—impossible for one person to monitor and comprehend what is important for the health of the product and business and customer experience.

The number of metrics is not as important as *what* is measured and monitored—and that harkens back to the business capabilities that are important to the customer. Customers may not pay much attention to a sign-up process if it goes smoothly, but if it is cumbersome, there are likely to be a lot of dropouts. Product leaders should consider which business capabilities provided to the customer are critical—those that, if resolved or improved, would have a positive impact on the business—and how the organization is performing against those.

Only that which is measured can be improved—so the first important step is to start measuring. From there, internal improvement goals can be set.

RESILIENCY

Within operations, an awareness of resiliency enables product leadership teams to monitor performance and make decisions. Resiliency is the philosophy of knowing that something is going to fail at some point—and there is no way to predict all of the many scenarios in which something may fail—so, assuming that to be true, what process will the organization follow to come back up as quickly as possible?

This is again a culture shift. Rather than asking whether a scenario is likely to occur, the product leadership team assumes that one will at some point and—ahead of that occurrence—ask what it means for the business and whether they have the right operations and technology to support that scenario.

The product leadership team needs to establish an operational resiliency program such that business, operations, and technology develop remediation guidelines on how to recover from these failures:

An insurance company may track standard operational metrics for the number of people logging in and the responsiveness of the platforms. Here, though, the head of product for an insurance company would want to also track the number of new policies signed today, how many people dropped out, how many new claims were filed, and the average claim size. Those metrics will give a good indication of company performance in terms of their business.

- What do you do if the website is down?

- What do you do if you are notified of a breach?

- What do you do if you have a software security issue?

Product leadership teams need to be able to question their organization, to assess what has to happen to come back into a normal state, how much time it will take, and what KPIs and service-level indicators will be missed because of it. The specific playbook developed will vary by industry and organization, but we want to highlight the importance of having the playbook at all.

The need for the level of specificity here comes down to the financial impact of the specific failure and what resiliency might entail. For some systems—an analytics data firm, for example—there is a higher threshold with a lower financial impact. The inability to make a trade on a trading application, on the other hand, can reflect a multimillion-dollar impact on a client. Healthcare and military systems also have a higher need for reliability in terms of being available.

In this level of maturity, the measurement of the previously discussed infrastructure and key business feature metrics will be most useful as a real-time dashboard. The data changes so frequently that it is more practical and functional displayed electronically rather than something like a business capability map, which can be printed out and displayed physically because it evolves significantly less often.

Product leaders need to establish and maintain this key

operational metrics dashboard as part of level one. Additional investments in operational metric collection and tracking need to be weighed carefully so teams don't go past the point of diminishing returns. Leaders may find it more prudent to make further investments in the product build process.

Level Two

BUILD

Level two asks, based on the bottlenecks shown in level one, where can the product leadership team bring improvements within the technology process? One such improvement may be continuous integration, wherein key elements of the process are automated as much as possible so that the teams spend less time on technology delivery and more time on building features. From a technology standpoint, we have seen that, generally, most bottlenecks occur in places where automation could be incorporated into the software development lifecycle in order to get features out faster.

Determining the amount of automation built into that lifecycle is a matter of asking focused questions. The three things a product leader should ask at this level are:

- *"How much automation do we have in terms of testing the features we're building?"*

- *"How can we automate the process of moving the change into an environment where we can validate whether or not this feature actually works?"*

- *"How can we scale all of these deployments so we can have this particular feature available to as many customers as possible, as reliably and as automated as we can?"*

Product leadership teams are better off investing in these core software development lifecycle automation steps; however, it is unlikely that an investment committee will give approval to focus solely on "fixing the factory," while adding no new features. Usually, the allotment will be divided, with the heaviest allotment of the budget going to new features and minor allocations toward technical debt and automation. If this is the case, how does a product leader make a choice on where to invest when it comes to fixing the factory? On continuous integration or test automation? Or should the focus perhaps instead be more on automated data creation needed for performance testing? We recommend pursuing the staggered approach here—take up development automation tasks that are most frequently required for a given product or service, amongst Build, Test, and Deployment Automation areas, that could have a greater impact for the entire product portfolio, increasing coverage of automation iteratively, and establishing a foundation for a mature development automation. We will expand on how to improve automation for the factory as part of level-three maturity.

Another bottleneck area to improve occurs when, historically, software deployments are a siloed responsibility across CIOs, CTOs, and data infrastructure teams. The infrastructure team works on their own schedule, with their own tools. The technology organization builds deployment components with their own tools. Automation is a cultural shift away from these silos and toward cross-functional product teams in order to deliver a better experience for the customer. During this level of maturity, the automation should break these silos, not just tooling with the product team. For example, allowing containerized web applications will simplify the CIO's task of setting up specific virtual machines (VMs) and preinstalling software.

Google is a big proponent of this culture of automation. A product owner assigns time and budget to automation, taking some money away from features and capabilities to do so. That lateral budgeting—being able to continuously monitor and fix those gaps—is built into the culture of that product team. Not only does that resolve some of the bottlenecks that arise, but it also helps prevent them in the future.

In this level, we also need to mature the dashboard metrics we discussed in level one to be more comprehensive yet not overburdening to the product leadership team. A set of key metrics is listed below as reference, which could be expanded upon.

We worked with a large digital transformation project for a private equity client who has purchased a portfolio of intellectual property data management firms. They were trying to build a new system that allows them to have faster onboarding of European and Chinese intellectual property, to become the "be-all, end-all" for intellectual property scanning and protection. It is a multiyear transformation to build this new platform, but in the weekly status report meetings, much of the energy is devoted to discussing features and functions, with very little bandwidth, if any at all, devoted to automation and development of the factory health. A leader might ask their team, "What can we be doing now to make things flow faster next spring?" It takes leadership to proactively make that discussion happen because it is very easy for features and

Metric	Undefined – L1	Defined – L3	Optimized – L5
Feature cycle time Time to take feature to production	>24 Weeks	4-12 Weeks	<2 Weeks
Number of releases Number of releases in a month/week/day	>12 Weeks	2-6 Weeks	<1 Week
Sprint predictability Number of committed stories delivered / Total number of committed stories	<25%	50-75%	100%
Functional test execution time Time taken to execute the testing	>4 Weeks	1-2 Weeks	< = 2 hours
Defect removal efficiency (DRE) Number of defects resolved by the development team/ total number of defects	<20%	40-60%	>80%
Code build time Time required to build the deployable packet	> 6 hrs	2-4 hrs	< 10 mins
Release and deploy time Time required to release to production	> 2 hrs	< 1 hr	< 10 mins
Time to create a new environment Time it takes to provision environments for developers	> 8 hrs	< 2 hrs	< 5 mins
Time to fix a broken build (Mean time to repair – MTTR) Time lag between broken (including compilation or quality gate failure) build to the fixed build	> 1 day	< 45 mins	< 10 mins
MTBF (mean time between failures) The up-time between two failures states of a repairable system during operation	< 1 day	< 2-4 weeks	< 24-48 weeks

Illustrative product engineering SDLC maturity levels

Figure 20: Directional maturity levels across key development operations metrics

functions to dominate the team's attention. With one client we worked with, we were able to introduce an automation mindset (and framework) into their backlog with help from the client leadership team, enabling us to improve the speed at which we built and deployed the software. We had to invest in people and time to bring this new capability to the team. Although it took away bandwidth from features in the short term, over the long term, automation enabled significantly faster daily build and testing processes.

OBSERVE

Level two in terms of production sees the product leadership team track and measure additional metrics: How are customers interacting with the product? What features are being used?

What activities do they do more often? Where do issues exist? What features are customers particularly interested in?

The maturity in this level is to move beyond key infrastructure and operational metrics to feature level observability. The ability to set up the observability platform to include attribution of user activity to features, revenue to features, and understanding proactively on a daily basis is what moves the needle from the product standpoint.

Many upstarts and technology-focused organizations make this level of maturity a basic building block, utilizing off-the-shelf tools such as Google Analytics and mobile analytical tools such as Google Firebase, Heap Analytics, etc. These product feature attribution platforms provide key insights to product leaderships on where to invest engineering dollars.

It is imperative that product leadership establish such tools to really get into the heart of the product in a production environment because it is the golden source of information.

More often than not, product leadership and operational teams have several hypotheses on consumer or market psyche but no way of testing the success of a feature through numerical analysis without product feature observability. Marketing teams need to know the marking attribution of campaigns and success of customer and revenue conversion, which also requires similar platform enablement.

Level Three

BUILD

Level three consists of applying and instrumenting the software build flow built in level two to upcoming product features:

- *How many features are we developing?*
- *Where are we with all these features? Which ones are in which stage of this automation?*
- *What percentage of this feature has been tested, if any?*
- *What defects are coming out? How many of those defects are being detected automatically?*
- *What kind of customer feedback are we receiving?*
- *What is the velocity of the team? How much progress are we making against that roadmap we had, now in a more automated and instrumented fashion?*

At this level, the product leadership team should have a comprehensive view of all critical features and capabilities that are in progress. It may not be possible to implement this level of visibility for every feature or capability in the backlog. But the aim should be to continually automate and optimize the flow, beginning with criticality. Instead of attempting to implement it across every product, a product leader should look at how much maturity they need for those products that have the most critical operations. For example, if a product under a portfolio leverages a multitude of frequent releases, it makes sense to invest time in deployment automation, versus a critical calculation engine, which requires a robust data setup and test automation functionality. Our recommended approach is described in the following visual.

Engineering teams should get comfortable with end-to-end automation as a philosophy first, before trying to increase coverage in all areas of services within a product. This builds confidence and economies of scale.

No one could possibly achieve 100 percent coverage; that

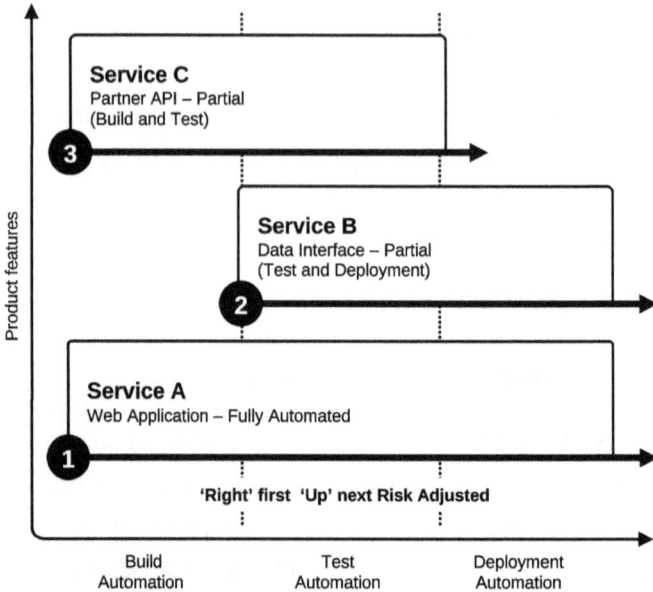

Figure 21: A sequential approach to increasing automation coverage across the product architecture

is too expensive. What is crucial, however, is to increase the end-to-end automation of services that are picked up first and then to incrementally add coverage (testing and automation patterns) toward 80 percent to 90 percent while adding more services, based on risk order of priority.

OBSERVE

As more metrics are added in production, maturity becomes a continuum where a product can be made more robust as it is made ready for production. Level three optimizes this process to better benefit both the business and the customer's experience. Not only are activities measured in terms of customer activity but also by feature activity, allowing product leaders to evaluate how to improve a particular feature or whether it continues to be necessary. The optimization process is a feedback loop wherein product leaders are not only looking at numbers but also taking action.

In level three of flow, product leaders can begin deploying A/B testing to get impactful real-world feedback: Is it really working? Is it scaling? Are people complaining? The ability to measure feature cycle times allows for optimization of the flow where bottlenecks occur.

A BCG study found that, by investing in tools and practices in software lifecycle automation and observability of the product, the organization was able to develop a sustained competitive advantage.[29] A quote from ING Bank's CIO as part of this study reiterates our principles around Agile and integrating Operations and Technology into it, stating, "We found that working in an Agile way solely in development didn't really make much of a difference. IT operations needed to be included as well, since that's basically where the buck stops before you go into production."

Building a factory that can release great products in an efficient way and measure the usage is a great benefit to product leadership teams. The decisions can be made based on source of truth data, which is production!

KEY TAKEAWAYS:

- The Build and Operational metrics used to measure product health are as important as business metrics.

- Product leadership needs to be cognizant of the amount of work and engineering that goes into software production operations.

- Software value streams are excellent tools to identify bottlenecks and measure improvements.

- Measure usage metrics from the customer point of view; production data is the source of truth.

- The ability to observe production and build a resilient product is critical in improving customer experience.

PRINCIPLE 6

KAIZEN

Relentless Improvement Is the Only Way to Push an Organization Past the Inertia of Complacency.

"More important than the actual improvements that individuals contribute, the true value of continuous improvement is in creating an atmosphere of continuous learning and an environment that not only accepts but embraces change."

—JEFFREY K. LIKER, *The Toyota Way: 14 Management Principles from the World's Greatest Manufacturer*

Born of Toyota, the Japanese term *kaizen*, which literally means continual improvement, has developed into a universally accepted methodology used in lean manufacturing to evaluate every part of a factory, making ongoing tweaks so that the whole is moving more effectively and expeditiously, not just the parts. Although kaizen began as a manufacturing construct, it has been lifted and shifted into a product development mindset and methodology.

In business, the fundamental philosophy of kaizen is the notion of instilling an organizational culture of continuous improvement—in all areas, but with a means to an end. Ultimately, all improvements made should be additive toward the end goal of the organization, whatever that company's mission statement or value creation process may be. Some people misinterpret kaizen to mean optimizing everything for the sole purpose of optimization for optimization's sake, but in fact kaizen should start with determining an organization's end goal and then examining what bottlenecks have become hindrances toward advancing their objectives. The continuous improvements of kaizen should begin wherever those bottlenecks are found.

Simply reading these principles and implementing the previous five principles once is not enough to help teams advance to their highest-performing state because it is human nature to revert to what is comfortable. Instituting the kaizen action into an organization via the Momentum Methodology detailed in this principle brings about change and sets into motion the continuous improvement that product leadership teams seek.

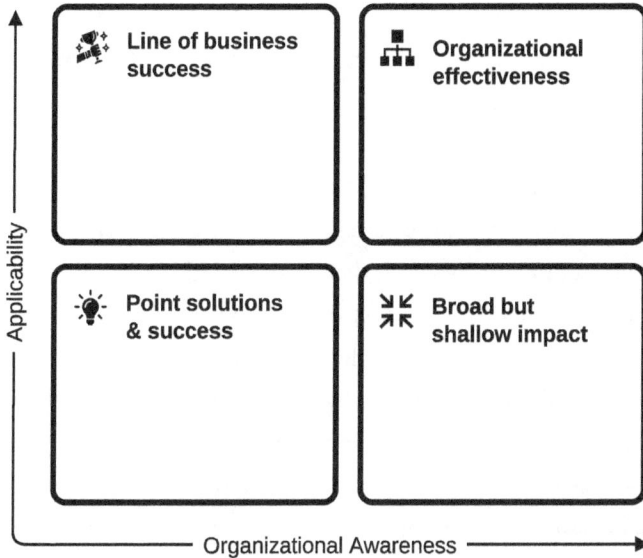

Figure 22: Benefits of applying Momentum principles through kaizen

KNOCKING DOWN
THE DOMINOES

The five principles presented thus far have the most impact for a product leadership team to improve the momentum of their organization. Although the maturity model for each principle has been presented linearly in this book, the process to advance through that maturity is not necessarily sequential.

Many product leadership teams will be tempted to tackle all five fronts simultaneously. That shotgun approach is less effective than taking the time to establish at least a rudimentary level in the first principle. To use dominoes parlance, start by playing the heaviest tiles first and laying the foundation on the business capabilities principle and the overall product strategy. The next priority point—be it re-architecting the product, restructuring the team, or investing in the infrastructure—will be a natural byproduct of that initial assessment.

It also important to highlight that improving the maturity on any single principle by itself will not add holistic value to the product or to the business; that is not what drives momentum. As an example, if a product leader overinvests analysis time toward business context value streams but has not realized any benefits of doing so—by re-architecting, re-operation-alizing, and/or changing the organization structure to affect it—that analysis alone is not going to have a business benefit. Let's assume that during the business context assessment, the product leader identifies a value stream requiring optimization: reducing the number of claims platforms to efficiently create a single organizational claims platform. The analysis performed at that third level of maturity of the claims process business context needs to be taken to the enterprise architecture team for re-architecture. From there, it might need to be taken to the operational team to identify the best way to implement those changes. These principles are independent in terms of what needs to be done, but the product or business will only realize the value if improvements are made relative to that business benefit across all principles.

The goal in the framework presented here is not to advance through the principles sequentially from one to five at level one, then start over and move through them again at level two; nor are we prescribing that it's necessary for an organization to achieve a specific level in each area. Product leaders have competing priorities, and teams will be pulled into other projects, completing other necessary tasks in order to keep the lights on. It is impractical to focus solely on change.

Instead, product leadership teams are asked to consider their specific business pain points and then determine how they can move up the maturity continuum in a particular dimension—with a specific reason for that advancement, not just moving up the maturity model for its own sake. The further an organization moves in the maturity model, the more effective that organization is, but trying to chew more than a team can swallow—or attempting to advance more than an organization can adapt to—is a recipe that ensures those change efforts will falter.

We consulted for a banking client who wanted to create a complete inventory of the process maps. It took months for the team to interview several internal stakeholders and move the process flows to level two. During those seven months of documentation, the core banking team who was dependent on this information moved on with their own priorities. Additionally, in selecting a vendor as part of their platform modernization efforts, the re-architecture that was triggered outdated the process maps that were created to begin with.

This is an example of an organization getting too far into the weeds on one principle without a clear north star, instead of making iterative progress on all principles tied toward meaningful outcomes for the product and its stakeholders (internal and external).

Because these principles are interlinked, and none of the attributes that make up the principles exist in isolation, they are not designed for product leadership teams to push further up the maturity curve on one principle without calibrating the others or moving them all along progressively. A person wanting to improve their health must examine and improve across multiple areas such as exercise, diet, and mental health; similarly, a combination of factors will assist an organization to mature holistically.

All these factors are considered simultaneously, but advancement is likely to occur at different rates in different areas. If significant improvement is necessary within business capability value streams and architecture, then data may not be a high priority. One product leader we worked with did not need to focus of the commercial benefits or consolidation of data, so Principle 4 was not top of mind for that client. Depending on the circumstances, one principle may be pushed a little further ahead than others at any given snapshot in time, but over the long graph, the totality needs to expand to achieve true effectiveness.

Additionally, as a team progresses to higher levels of maturity, each level can take exponentially more effort to complete because of the additional due diligence and skillset needs required. There comes a point of diminishing returns beyond which it does not make sense to continue advancing on a specific principle in the business context of the product. Boiling the ocean on business context, for example, will lead to opportunity cost in another area.

THE MOMENTUM METHODOLOGY AS A GUIDING PRINCIPLE

The demands of running a product engineering team are many, and the number of meetings, cross-functional teams, moving parts, and buy-in required to address change can bog down any product leadership team. Be it addressing each principle or just trying to move from one level of maturity into another, there is always the possibility of overwhelm. To address this, we have created the Momentum Methodology, which determines directionally how to calibrate the speed at which a product team progresses on each of the first five principles.

We were intentional in naming these meetings momentum meetings. While other forums and recurring meetings may exist that approximate what these meetings may be called, we believe it is important to have meetings that deliberately focus on the act of improving across the dimensions critical to the product team.

The Momentum Methodology is as follows:

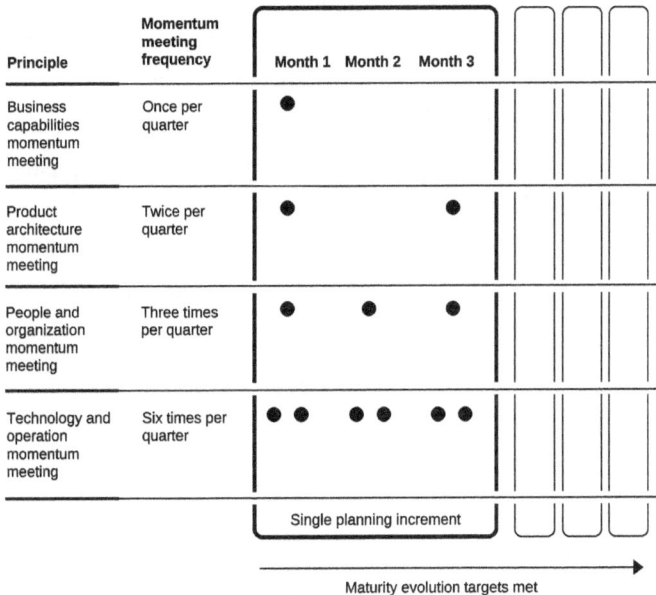

Principle	Momentum meeting frequency	Month 1	Month 2	Month 3			
Business capabilities momentum meeting	Once per quarter	●					
Product architecture momentum meeting	Twice per quarter	●		●			
People and organization momentum meeting	Three times per quarter	●	●	●			
Technology and operation momentum meeting	Six times per quarter	● ●	● ●	● ●			
		Single planning increment					

Maturity evolution targets met →

Figure 23: The Momentum Methodology

This methodology lays out the minimum number of meetings a product leadership team should have per quarter to move the needle on each principle in an impactful manner. Our Momentum Formula is organized to correspond with most Agile frameworks, which construct delivery cycles in quarterly increments. From here, each team will have to adjust the specifics of the formula in the new direction.

> What can vary from team to team is whether to have distinct meetings for each or, if the team is small enough, to combine them with existing meetings, making sure these topics become part of those agendas.

While the product leadership team can assign an individual to research the current state of the business upon which to build the principles, a number of team meetings must take place to enact a plan. Each momentum meeting is designed to have a precise spirit and outcome, addressing specific capabilities, moving up the maturity model, and evolving what appears on the wall. Although the product leadership team may have to respond to certain triggers that are more reactive, the intent with this formula is to move these meetings toward being proactive for the five principles. Each momentum meeting should be used to remind the team of the principle's relevance, re-benchmark as a team, set goals of how to mature within this principle in the coming quarter, and then reflect on the progress made in the previous quarter. Across each principle, there is a standard objective to continue iterating and improving, expanding the impact across dimensions as the team moves up the maturity levels.

A product leader sets up these meetings to be purposeful; thus, they want to ensure they have the correct objectives going into a momentum meeting and actionable quarterly deliverables coming out of it. Before scheduling each momentum meeting, they should ask:

- *Why am I having this momentum meeting?*
- *What objectives do I have?*
- *What is the level of maturity we desire?*
- *What are the accidental triggers?*

Planning this agenda ahead of the meeting allows participants to focus on the specific goals of the meeting. Obviously, the specific agenda each quarter will be situational to each organization, but the practice of using prompts to determine the agenda is useful for all product leaders. With the tone and agenda of these meetings preset, everyone attending will be ready to discuss the topic underpinning it all: continuous improvement.

Many organizations use frameworks such as PDCA (Plan, Do, Check, Act) or OODA (Observe, Orient, Decide, Act) loops as a framework for ensuring improvement occurs. If we take John Boyd's OODA loop as an example:

- **OBSERVE:** *The first step is to observe and understand the exact level of maturity the team is exhibiting in the context of a momentum focus area.*

- **ORIENT:** *Analysis and synthesis of observations allow the product leadership team to determine appropriate and effective corrective measures or improvements that need to be made to achieve the desired outcome, as well as ensure that such action is moving toward a longer-term north star for the team.*

- **DECIDE:** *Given a multitude of options and corrective paths, it is incumbent that the team and its leadership are able to arrive at decisions on how to move forward. For many organizations, this step tends to be where bottlenecks occur. Fear of making an incorrect decision must not paralyze a team from making a decision with the best available hypothesis.*

- **ACT:** *Put into action the decisions made without delay. A body in motion stays in motion, and the same is true when creating a culture of continuous improvement. Raising standards and acting to meet them is a discipline that must be ingrained into the fabric of the team.*

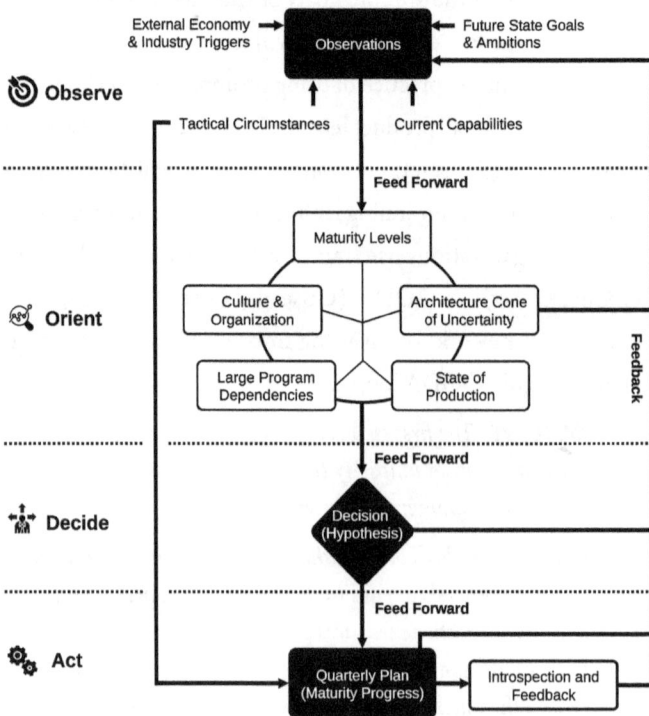

Figure 24: An adaptation of the OODA loop to introspect and advance across maturity areas of the Momentum Framework.[30]

When determining where to focus, prioritize areas that can be measured so that the outcome of an OODA/PDCA loop improvement can be quantified. Metrics play an important role in ensuring meaningful conversations in any momentum meeting and help to keep leadership teams grounded in quantifiable improvement targets. Throughout this book, we've called out metrics specific to each momentum area. In addition to principle-specific metrics, it is also important that the production leadership team keep top of mind more holistic metrics. As an example, %C&A (percent complete and accurate) is an important metric in value stream improvement to track that quality conditions are met at each stage of the value stream. It is not uncommon in interviewing software development teams to find that they have a %C&A of zero—meaning that input they receive often has to be reworked 100 percent of the time. More holistically, the Rolled %C&A (compounded effect of quality across the entire value stream) is less than 50, meaning 50 percent of the work has to be reworked at some point across the value stream.

Now let us look at each momentum meeting's frequency and focus.

Business Capabilities Momentum Meeting: Once per Quarter

One meeting on business context should happen once per quarter. This doesn't replace other meetings the product leadership team might have; this is a very specific meeting with product leads about the business context principle.

In Principle 1, product leaders have benchmarked their

portfolio business capabilities at a specific level. The question now becomes where to perform a deep dive, where to advance and make investments in this quarter or the upcoming quarterly cycle. It is impractical to consider moving an entire organization's business capability map to level four, so one goal of this meeting is to determine those areas necessitating the deeper focus, collectively with architects and business strategists. Sometimes a set of triggers—a regulatory change, market shift, customer feedback, or change in consumer behavior—will force the product leader to focus on improving a particular area of their business.

Outside of that, a strategic lens can be applied, allowing the product leader to prioritize specific capabilities to advance the level of depth they have on the capability maps of those areas and/or build deeper value streams in those areas. Those specific areas of focus, whether triggered externally or internally, then feed into the subsequent meetings that are going to take place and set the tone for the rest of the quarter. The business context meeting is the lead domino to then determine what else needs to change or improve down the line in the rest of the principles.

In the next quarter's business context meeting, attendees will touch back on those capabilities that were set as goals from the previous meeting and monitor the progress made on those before moving forward with new ones. These meetings should include reflection on what was previously committed to as well as the achievements derived from that commitment. This reflection is a fundamental part of the kaizen philosophy because it is a key factor in how organizations move forward.

For example, if in quarter one, a team has a business capability map with everything at a level one, the team will

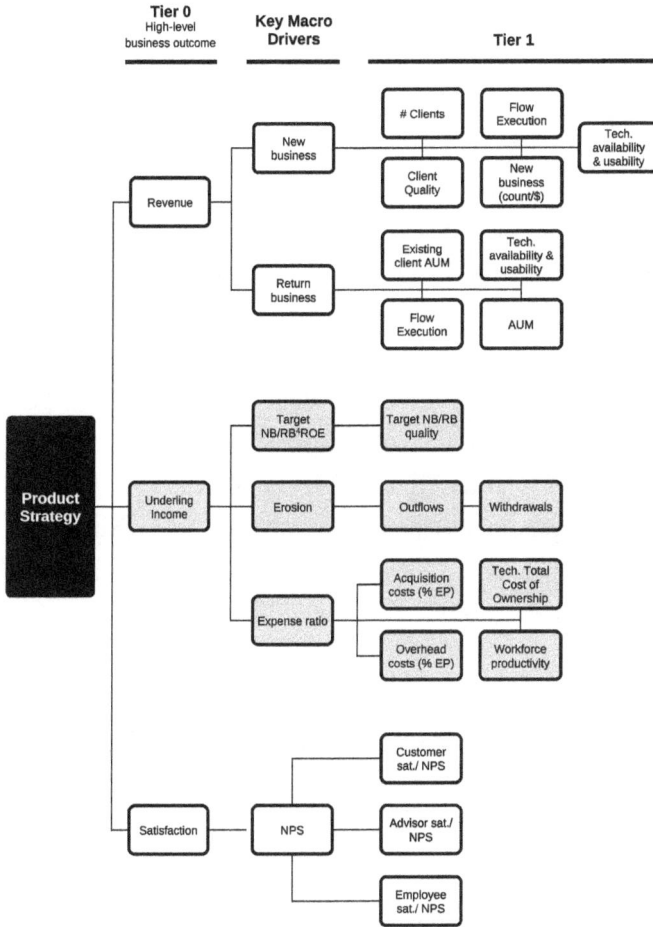

| Tier 0 High-level business outcome | Key Macro Drivers | Tier 1 |

Example KPIs to evaluate product management decisions at a wealth manager

Figure 25: Sample product strategy KPIs for a wealth management product

determine which portion of their product portfolio will be moved to the next level. When they return to the war room the following quarter, there should be a new plot of the business capability map on the wall with that capability or area fleshed out.

The business capability map depth should mirror the product strategy and the prioritized KPIs. As an example, the following diagram shows possible business strategy metrics for an insurance product. As the team prioritizes focus areas of improvement each year and quarter, the product capability map depth should mirror that same focus. If the team is focused on increasing renewals by 3 percent, the product team should prioritize going deeper on capabilities that directly impact the renewals process. This could include leveraging customer data to provide more predictive recommendations for their renewals, reducing friction points across renewal channels, as well as better tools for agents to help them manage renewals across their practice.

Enterprise Architecture and Data Momentum Meeting: Twice per Quarter

Two meetings per quarter can cover enterprise architecture and data together.

In the business context meeting, a decision was made as to which capabilities or sub-capabilities to drill down to further levels of detail, whether because they need improvement due to a pain point previously identified or because the organization needs to build a new capability. In this meeting, the architecture and data teams will overlay their maturity evolution on those capabilities that are important to the business, documenting the capabilities from an enterprise architecture and data standpoint, and creating and presenting artifacts to help the product leadership team make a decision in terms of the direction to take toward these capabilities. They present the new architecture and data plan for that landscape and the impact of the organization from a product standpoint so the product leader can determine whether it is practical and achievable.

Enterprise Architecture and Data Management artifacts require objective setting at the beginning of the quarter and assessment and executive readout toward the end of the quarter. This format focuses on what needs to be done, not only to document the capabilities to improve the level of maturity, but also to support the product leader in moving forward and making the change effectively and efficiently. These meetings take place twice per quarter because, while they are not as high level as the initial business context meeting, they are more detailed and require follow-on assessment projects to be completed. They must meet multiple times in order to ensure that the objectives are met before the teams meets again toward the end of the quarter from a momentum meetings perspective.

To provide a typical example of the outcomes of these activities for Architecture and Data, one of our clients determined, through market research, that they were spending too much money on reporting compared to others in their industry. Thus, they set reporting as the capability to go deeper on in the first business context meeting. Every product lead determined all the places they do reporting, why they do it, and what technologies they use so the head of product could build the system context diagram and the data lineage, level-three maturity for architecture and data, just on that reporting capability. This is a business-capability-led assessment, classic fit to improve the maturity of architecture and data principles while focusing on improving business capabilities.

The assessment for future state reporting architecture and associated deliverables were the focus of the quarter, and the final assessment and application rationalization to consolidate into a single platform was presented toward the end of the quarter. Part of the ongoing updates to capabilities and to the operations will be how our clients were doing against the reporting product they are going to rebuild.

People and Organization Momentum Meeting: Three Times per Quarter

There are three meetings per quarter to discuss people and organization.

The product leadership team has decided on a set of business priorities, and they've laid down the architectural and data foundation they'll need to enable those capabilities. Now, they will look at which teams will perform the necessary work, considering the health of those teams, and how the product leaders can move those teams up the maturity model in the zones where they want to invest. The product leader may ask, "What are the iterative improvements to make in order to help these teams have better delivery feedback, improve their velocity, or have a tighter delivery methodology?"

Most teams have varying sprint cycles—very advanced teams will do two-week sprint cycles, some teams do three, but most teams do four-week cycles, so the monthly cadence lines up well. Meeting once a month allows the product leader to collect the necessary data about the teams' performance, the issues they ran into, and areas for improvement to move them further along on the maturity curve.

Category	Metric	Definition	Insight
Productivity	Team velocity	Number of story points finished per sprint	While Velocity's primary use is to help teams judge their capacity when planning, it also reflects the productivity trend over time
Customer value focus	Time elapsed since Last Customer Demo	Number of days since last demo to customer (can be internal)	Monitors how often teams seek and receive customer feedback, how iterative their approach is, and the degree to which they embrace the MVP mindset
Product quality	Defects (software delivery) or other quality metric	Relevant metric that determines quality of delivery	Monitors the quality of work being done. Combine with view of delivery for insight into efficiency
Predictability	Story points Completed vs committed Trend	Number of story points that are completed vs. planned for in each sprint	Assesses reliability of planning capabilities
	Sprint Burndown	Amount of work (as measured in story points) remaining throughout a sprint	Assess flow of work over the course of a sprint and gauge likelihood of successful completion of committed work
Team engagement	Monthly survey	Pulse check measured against a set of Agile metrics (e.g., Agile benefits, team performance, Individual empowerment)	Assesses teams' perception of Agile enablement and empowerment
Agile maturity	Coach's team maturity Assessment	Health check filled out by Agile coach	Provides insight into the maturity of Agile practices adopted by teams

Squad health metrics

Figure 26: People or squad metrics for constant evaluation and trends

Operations and Technology Momentum Meeting: Six Times per Quarter

Finally, there are six meetings per quarter to discuss occurrences in operations and technology.

The tone of these meetings is generally twofold: first, what's happening out in the wild? Applications have gone live and there may have been defect issues in production, which indicate areas needing improvement or advancement prior to production. Second is the internal side of the house, which involves evaluating the overall infrastructure, tools, and processes to manage the rate of flow. What can be done to tweak and improve the organization in order to have a faster cycle time in a very healthy state?

These meetings occur fortnightly and should alternate their focus between what a product leader needs to accomplish in that business-as-usual setting and the change management or new projects coming up.

In the business-as-usual setting, the main purpose and objectives the product leader addresses are:

- *How are we doing in production?*
- *What are some of the production health issues we have been noticing with our products?*
- *Are people using features the way we expect them to?*
- *What insights can we learn from our product analytics to apply to our product?*

With regard to change management, the product leader needs to have visibility in terms of:

- *Where are we with respect to our plans? Is the velocity relative to Program Increment on track? Can we meet the next milestone release?*
- *Are we ready to move forward with this new feature?*
- *Are we going to move this feature into production in the next weeks?*
- *What challenges exist in terms of putting features into production? (Infrastructure scaling, vendor agreements, etc.)*

We had a meeting with a head of product who wanted to improve the DevOps maturity of max implementation, focused on improving the factory, flow rate, and automation coverage as discussed in Principle 5. We benchmarked where they were in the maturity model and set goals with them.

Coming out of that, the client realized that they had veered significantly off course due to a lack of an ongoing check-in on this topic, leading to the point where things had deteriorated to the point that they required a partner to help remedy the situation. This was not an instance of lack of knowledge within the client team; it was a case where lack of intentionality and consistency in focusing on this momentum area resulted in a degradation of a key function of the software development process. In contrast, had this team prioritized and ensured a Tech plus Ops momentum meeting was occurring at the frequency we have outlined, the feedback loop to the team would have minimized how far they strayed off course against their own desired performance.

After the first quarter with this formula of twelve meetings in place, the momentum for continuous improvement—kaizen—begins. During the first meeting of the second quarter the team examines accomplishments of the prior quarter and defines a different set of capabilities or sub-capabilities to focus on for the upcoming quarter.

COMMON CHALLENGES

Implementing a continual improvement culture does not come without challenges.

The most common challenge we see occurs when product leadership teams in large organizations want to know how **they compare to their industry peers**, or they ask for an average of all of their peers within the sector. That is generally not a harmonized number, but some clients leveraged industry consortiums to see where they would fall in comparison with other companies in their industry. These benchmarks—for example, data governance rankings in financial services industry—should be used solely as guideposts.

While it is natural to draw comparisons, the challenge comes in fitting one organization's challenges against another's—those challenges are not likely to be the same, so a comparison of each company's progress is irrelevant. Instead of trying to advance on all levels across all principles to maintain pace with other organizations, product leadership teams will be more successful determining what is going to move the needle most for *their* organization, relative to the kaizen principle, and then focusing on that.

Another challenge occurs when product leaders look at the template and **confuse these meetings with existing business-as-usual meetings that sound similar**. They may ask, "We're already having those meetings—why don't I have momentum?" This formula is not just about having the meetings; it's about kaizen, evaluating maturity, and using the meetings to push the maturity of each of those principles continuously. These twelve meetings focus exclusively on the topic of momentum and helping product leadership teams mature their organizations, with a product-specific or value-stream-specific view. The Momentum Methodology forces heads of product to shift from looking at each area in a silo to a more holistic view. It isn't designed to replace existing meetings; the intent is to have a meeting focused on a principle, with a specific agenda and goals for specific improvement outcomes.

Product leaders may also fail to find success if they are

not deliberate about adopting the kaizen philosophy of asking, "What is our evolution? What is a small improvement we're making this go-around?" Tied into that issue is failing to appropriately utilize metrics to measure impact and the change in momentum.

A similar challenge comes from **overly indexing on proxy metrics**, which tend to be internal metrics like team velocity, without taking into context the ultimate metric: how this value stream or product is supposed to create value. That metric can be measured by evaluating:

- *Is the product team generating more revenue, more profit?*
- *Is the product team seeing less churn?*
- *How is the team measuring the ultimate value of this value stream, and is that visible to everyone on the team?*

As an example, for Principle 5, the key argument to be made here is to include the engineering team Net Promoter Score (NPS) as proxy to gauge engineering productivity. In order to retain valuable engineers, the majority of product engineering organizations want to make sure those engineers are happy. Often one of the biggest contributors to low NPS amongst top-performing engineers is working in an environment where it is too difficult or time consuming to release new capabilities. Addressing those frustrations and making the investments that help engineers be more productive can drive up satisfaction and engagement.

Finally, **a lack of executive support is fairly common**. Sometimes change efforts (Agile organization change programs, digital transformation programs) are performed by a central change organization that is funded separately.

The product teams aren't necessarily bought in as they have a primary focus on product-specific objectives. Additionally, executives are more focused on the leading indicators versus lagging indicators. There has to be a focus on new capabilities and continual improvement.

A successful product leader learns to demonstrate small wins to their leaders—after all, success begets more success. Connecting back to metrics that affect the business as a whole increases the likelihood of gaining executive support.

LOOKING AHEAD: REACTIVE VS. PROACTIVE

There's always a reactive component that triggers an assessment—but there can also be a proactive component, wherein the product leadership team examines each of these areas *before* an issue crops up to ask, "How can we evolve and innovate to make our factory better?"

In a reactionary mode, the product leader pushes on each of these principles as far as they can in the context of the immediate business need. When acting proactively, however, it is possible to embed a review—called a PI Planning event in Agile—wherein nothing is immediately on the radar, but the team is pulled together to take a step back, look at the product or portfolio for which they are performing this PI Planning, and walk through the principles to see what level the team has achieved for each:

- *Do they have a solid understanding of their business capabilities?*
- *Is this an opportunity to inch up the maturity curve?*

- *How are they doing on the architecture?*
- *How are their data management discipline and practices helping the product roadmap?*
- *Do they have any missed opportunities from a business perspective relative to the industry?*
- *Are there any upcoming risks called out from a strategy point of view?*
- *How are these Momentum principles operationalized within the organization?*

There are a ton of meetings happening in a product leader's world, but this is the stake in the ground to get people to pause and focus and introspect on the specific topic of improvement in these five principles.

Ultimately, moving through this framework is never truly complete. Once a goal is hit in one area, a new issue that pops up in another area or a new focus for improvement will be found. Additionally, level three for each principle is never-ending; it will never be completed because it represents kaizen: ongoing, continuous improvement.

KEY TAKEAWAYS

- Continuous maturity across all five principles requires a sixth principle to bind them together: kaizen

- In large organizations, it is easy to lose momentum if not properly followed through, due to shift in focus, reorganization, changing priorities, etc.

- All five principles are key to achieve momentum in building a strong product culture, architecture, data assets, and domain-rich features

- Follow our Momentum Formula and tailor it to the size of the products/teams

- Show metric-based improvements that are tied to business outcomes

CONCLUSION

As we conclude our approach to effectively implement the principles that help drive momentum, we return to the three colleagues we started our book with: a product manager, an architect, and an engineering team lead at a large financial services enterprise.

A year after implementing the Momentum Framework with their teams, they find themselves in a different place than when they started.

There is a sense of calm within the team. The framework helped to remove the circular ambiguity that the trio found themselves deadlocked with initially. There is hopefulness and clarity in the planned maturity evolution of the team.

The structure has helped frame product investment conversations holistically beyond technical features to the infrastructure and organizational improvements that allowed their teams to have a business impact.

Like a rowing team in sync, the processes and methods the team uses to plan, organize, and deliver their work are optimized for what the group and its customers need. Morale is up, having unlocked some quick wins that have already been delivered this year.

It wasn't all roses, though. The leadership learned the hard way during the initial stages of the rollout that implementing a framework required more than learning and disseminating the knowledge about the framework. Beyond the knowledge, it also demanded that the team be effective—effective in implementing the framework, effective in staying the course through the messy middle, and effective in managing fears, uncertainty, and doubts through the process.

EFFECTIVE LEADERSHIP

Thus far, we have examined the principles that drive momentum, as well as the necessary structure needed for product engineering teams to maintain and accelerate that momentum. It is an important job of the leadership team to keep in front of the product team a holistic view of what momentum means for the team and its customers. As we close out this book, we also want to reflect on the qualities that a leader needs to embody to effect this change. A key part of that is the realization that it is not enough to be right; a leader must also be *effective*.

What does effectiveness mean in this context?

Effectiveness means that a leader can achieve the desired outcome. It is a level past righteousness, which focuses more on selecting facts to prove a point, demonstrating expertise, and attempting to win over opinion. In and of themselves, none of these are guaranteed to result in positive traction toward the intended outcome.

All this is done in favor of forward movement and progress for the team.

Inspired by executives we have had the opportunity to

work with over the years, we've pulled together a guide to help leaders be more effective on their journeys to infuse a momentum culture into their product engineering teams.

Allocate Time to Prioritize Momentum Progress

Effective leaders make time for planning and reflection. As trivial as this may seem at face value, it is an easily missed step to being effective. It is important to find the space to look at themselves in a detached manner and appraise where exactly their efforts might be blocked or curtailed.

It is important for leaders to protect time on their calendars regularly to prioritize learnings from prior momentum meetings and examine where the bar needs to be raised or adjustments need to be made. This is a difficult balancing act. Leaders need to be able to work on momentum initiatives outside of these meetings, but they will also want the time before those meetings for preparation and reflection. Balance is key here.

It is important for leaders to keep a record of how much time they allocate toward this activity each week. Effective executives manage their time optimally by constantly keeping a log, reviewing it periodically, and setting deadlines for important activities based on their judgment.

Know Where to Contribute

The effective executive focuses on contribution. He asks, "What can I contribute that will significantly affect the performance and the results of the institution I serve?" This is the central question in leadership because the emphasis on responsibility sets the tone of accountability across the product team.

As leaders work through the principles outlined in this book, their experiences and expertise may lend to taking greater ownership in some areas. Conversely, being cognizant of areas where they have weaknesses serves to ensure that those leaders have appropriately assigned talent to lead the implementation and application of these principles.

The effective executive understands that he is a repository of his own experience and expertise. He understands his decision-making process, his limitations, and his potential for growth and development. By continuing to learn how to be a better leader, he continuously refines those elements, which he can bring to bear in service of the institution.

Build On Teams' Strengths to Start

As an extension of the above, every team also has a unique slant that may allow it to excel in one principle versus another. Focusing energy on areas where the team already has strengths allows the team to reap benefit earlier as well as move with the psychological winds of victory in their sails.

Concentrate efforts where meaningful progress can be made. Focus is the secret ingredient to helping a team on their path to being more effective.

The early wins enabled by this focus will help to get the necessary momentum to work through other more difficult areas.

Be Decisive

A key responsibility that will fall on the shoulders of any leader is the burden of decision-making.

The very nature of the principles will dictate that there will be multiple paths and approaches the team can take at any moment. It is imperative that the product engineering leader be decisive in the path forward. These decisions, in some instances, will be a judgment based on experience and principle; in other instances, they will be made pragmatically, based upon the merits of the case.

As Marty Cagan outlines in his book, *Inspired*, shifting priorities is a significant contributor to product team churn and substantially reduces total throughput and morale. The lack of courage to be decisive and reliance on the seemingly well-intentioned safety of consensus will mean that decisions are very hard to make and everything slows to a crawl.

Hard Work Beats Talent

As high school basketball coach Tim Notke says, "Hard work beats talent when talent fails to work hard." Heads of product already possess the talent they need to do their jobs well; they've simply been lacking the structure, framing, and a proven approach to apply to their teams. Effective leadership is not a mythical concept that only a select few have the skillset to accomplish; with the right tools, hard work, and the correct approach, *anyone* can do it. By laying the groundwork established here, implementing this methodology, and following through on it methodically, our framework will not remove

BEING EFFECTIVE:
A NAVY SEAL'S PERSPECTIVE

"There's only two measures that matter, and that's effective and ineffective. Are you effective in accomplishing the mission? And if you are, then you've got to find a way to become even more effective. If you are ineffective and you are not accomplishing the mission, then you have to take ownership of that and solve that problem. If your team is not getting the job done, as a leader, you have to drive those standards.

"I learned pretty early on that in the SEAL Teams there's some strong personalities there, which is awesome. That's one of the great things about the SEAL Teams, and there's always this tension between "We should do it this way," or "We should do it this way," "We should train this harder," "We should focus here or there." But as a leader, you have got to maintain the standards, and there are standards that just cannot be compromised. You've got to push hard; you've got to drive that performance. You've

every roadblock a head of product is likely to encounter—but it will circumvent the larger obstacles likely to derail a team or organization, leaving the path clear for forward momentum.

A CTO we worked with said, "I wish I'd had this book sitting on my desk when I was wrangling with my CTO position, just to be able to refer back to this north star and keep us on the right track."

Sustaining change demands a high degree of endurance and perseverance from its leaders. Aspiring product leaders are likely to have obtained this book because they want to successfully transform their product teams. It is a common perception got to set the bar for that, and if you don't do that, no one's going to do it. It's the performance that you see, and if it's substandard, you got to push.

"Again and again, you must push and set that bar high. Now that being said, you can't drive your team to the ground; you have to lead them. You can't be a slave driver. You can't destroy your team. You can't be overbearing. But—for those things that really matter— push performance to the next level. You have got to set that standard, and I think that's what makes truly the best military units and the best teams out there great.

"It's not the words that you say. It's not the email that you send. It's not the banner that you created on the wall or the PowerPoint slides that you flip through.

"It's not what you preach, it's what you tolerate."

—Leif Babin, US Navy Seals Platoon Commander

that it takes an inordinate amount of talent to achieve the types of the transformation they envision for their teams. However, we find time and time again that as important as talent may be, the consistent execution of the framework is more of a deciding factor in whether a team transforms.

WHY DO ALL THIS?

If a product leader were to step back and ask, "Why should I go through all the effort and grind to make sure I have the best product team?" they only need to let history be their guide.

For large organizations, specifically S&P 500 companies from 1957 to 2003 (prior to the dot-com era, mostly industrials and commodity companies), the opportunity existed to not only survive but also reposition themselves as thought leaders. In fact, research compiled by the Institute for Management Development (IMD) notes that the lifespan of these big companies was down from sixty-one years to eighteen![31] This is not a great percentage, but there is a chance.

This symptom of becoming irrelevant, described as "organizational inertia,"[32] requires constant feeding and caring of organizational strategy.[33] It is difficult to achieve transformations within large organizations given change of pace of technology and customer expectations. Our principles and framework discuss how to *build momentum at these large organizations,* not just avoid that inertia.

Product leaders know they have the talent or can acquire talent, but energies needed to be channeled. Businesses are always competing to become digital and provide frictionless customer experience. Regulators are always playing catch-up with new rules and regulations to reduce systemic risk, improve privacy, and level the playing field between David and Goliath.

In our combined forty years of interacting with various technology and business stakeholders at large organizations, whatever problem we are asked to solve and no matter the situation we were put in, our approach—even before we ever called it the "Momentum Framework"—always gravitated toward leading with one or more of these principles.

Our principles, whatever form they may take in a specific organization, are universal truths: be customer obsessed, aggressively remove legacy, build great architectures, focus on data, organize a powerful team, and consistently level up your processes and activities.

It's our hope that this book has given you a solid starting point to begin reaping the rewards that come from being on a high-momentum team. We wanted to deepen your understanding and give new insights into how these principles feed off each other in order drive performance. But, as with any learning process, action precedes results! So act.

Get started today by following the principles we outlined for success; may all of them aid you to work toward achieving mastery at managing momentum within yourself and others around you.

FURTHER READING

Berson, Alex and Larry Dubov. *Master Data Management and Data Governance*. New York City: McGraw-Hill Education, 2010.

Bossidy, Larry, et al. *Execution: The Discipline of Getting Things Done*. New York City: Random House Business Books, 2011.

Cagan, Marty. *Inspired: How to Create Tech Products Customers Love (Silicon Valley Product Group)*. Hoboken: Wiley, 2017.

Drucker, Peter. *The Effective Executive: The Definitive Guide to Getting the Right Things Done (Harperbusiness Essentials)*. New York City: Harper Business, 2006.

Frederick Brooks Jr., *The Mythical Man-Month: Essays on Software Engineering*. Boston: Addison-Wesley, 1995.

Fowler, Martin. *Patterns of Enterprise Application Architecture*. Boston: Addison-Wesley, 2002.

Kersten, Mik. *Project to Product: How to Survive and Thrive in the Age of Digital Disruption with the Flow Framework*. Portland: IT Revolution Press, 2018.

Martin, Karen, and Mike Osterling. *Value Stream Mapping: How to Visualize Work and Align Leadership for Organizational Transformation*. New York City: McGraw-Hill Education, 2013.

Peter Thiel and Blake Masters, *Zero to One: Notes on Startups, or How to Build the Future*. New York City: Crown Business 2014.

Pöppelbuss, Pöppelbuß, Jens and Maximilian Röglinger. "What Makes a Useful Maturity Model? A Framework of General Design Principles for Maturity Models and Its Demonstration in Business Process Management." ECIS (2011).

Robert C. Martin. *Clean Code: A Handbook of Agile Software Craftsmanship*. Hoboken: Prentice Hall/Pearson, 2008.

ABOUT THE AUTHORS

Software engineers and organizational change experts **SHIVA NADARAJAH** and **SURESH KANDULA** help Fortune 100 enterprises modernize their infrastructure and enhance product capabilities. Shiva studies how product organizations perform, evolve, and compete in a digital world, helping large firms establish themselves and organize their teams. An engineer at heart, Suresh specializes in advancing enterprise architecture and platform development principles within the engineering community. For more information and additional resources, visit ArtofMomentum.com.

ENDNOTES

1   Klemens Hjartar, et al., "Next-Gen Technology Transformation in Financial Services," McKinsey & Company, April 2020, www.mckinsey. com/~/media/McKinsey/Industries/Financial%20Services/Our%20 Insights/Next-gen%20technology%20transformation%20in%20financial%20services/Next-gen-technology-transformation-in-financial-services.pdf.

2   Shital Chheda, Ewan Duncan, and Stefan Roggenhofer, "Putting Customer Experience at the Heart of Next-Generation Operating Models," McKinsey & Company, March 17, 2017, www.mckinsey.com/ business-functions/mckinsey-digital/our-insights/putting-customer-experience-at-the-heart-of-next-generation-operating-models.

3   Wendy Grad, John Grudnowski, and Sarah Dey Burton, "The Measurement Advantage." Bain & Company, March 26, 2019, www.bain. com/insights/the-measurement-advantage.

4   "Business Architecture." Wikipedia.org, accessed June 1, 2021, en.wikipedia.org/wiki/Business_architecture.

5   Robert Charette, "Inside the Hidden World of Legacy IT Systems." IEEE Spectrum, July 28, 2021, spectrum.ieee.org/ inside-hidden-world-legacy-it-systems.

6   "Introduction to Building Blocks." The Open Group, 1999, www.opengroup.org/public/arch/p4/bbs/bbs_intro.htm.

7   Grad, et al., "The Measurement Advantage."

8   "Introduction to Building Blocks."

9   "Unlocking Success in Digital Transformations," McKinsey & Company, October 29, 2018, www.mckinsey.com/business-functions/ people-and-organizational-performance/our-insights/unlocking-success-in-digital-transformations.

10 John Bottega, "EDM Council DCAM and Analytics Update," EDM Council, March 2020, cdn.ymaws.com/edmcouncil.org/resource/ resmgr/featured_documents/dcam_analytics_mb_0320.pdf.

11 "Growth Readiness Study," State Street, October 2020, www.statestreet. com/content/dam/statestreet/documents/Articles/454_State_ Street_Growth_Readiness_Study_2020.pdf.

12 Paolo Benedet, et al., "Analytics-to-Value: Digital Analytics Optimizing Products and Portfolios," McKinsey & Company, April 21, 2021, www.mckinsey.com/business-functions/operations/our-insights/ analytics-to-value-digital-analytics-optimizing-products-and-portfolios.

13 Jeff John Roberts and David Z. Morris, "Robinhood Makes Millions Selling Your Stock Trades...Is That so Wrong?" Fortune, July 8, 2020, fortune.com/2020/07/08/robinhood-makes-millions-selling-your-stock-trades-is-that-so-wrong.

14 "California Consumer Privacy Act (CCPA)," State of California Department of Justice—Office of the Attorney General, access year 2020, oag.ca.gov/privacy/ccpa.

15 Bryan Petzold, Matthias Roggendorf, Kayvaun Rowshankish, and Christoph Sporleder, "Designing Data Governance That Delivers Value." McKinsey & Company, June 26, 2020, www.mckinsey.com/business-functions/mckinsey-digital/our-insights/designing-data-governance-that-delivers-value?cid=app#.

16 Bottega, "EDM Council DCAM and Analytics Update."

17 "15th State of Agile Report," Digital.Ai, 2020, digital.ai/resource-center/analyst-reports/state-of-agile-report.

18 Phil Maddaloni, "Agile at Scale: Comparing the Popular Scaling Frameworks," 6kites, May 27, 2019, www.6kites.com/blog/comparison-scaling-agile-frameworks.

19 "Scaling Agile Survey 2017," cprime, accessed January 21, 2021, http://www.cprime.com/resource/white-papers/scaling-agile-survey-2017; "2nd Annual Agile at Scale Report 2019," cprime, accessed October 6, 2020, http://www.cprime.com/resource/white-papers/2nd-annual-agile-at-scale-report-2018; "3rd Annual Agile At Scale Report," cprime, 2020, www.cprime.com/wp-content/uploads/2021/01/Cprime-Agile-at-Scale-2020-Final.pdf.

20 Darrell Rigby, Jeff Sutherland, and Andy Noble, "Agile at Scale," Harvard Business Review, May–June, 2018, hbr.org/2018/05/agile-at-scale.

21 "Program Evaluation and Review Technique," Wikipedia.org, accessed May 4, 2021, en.wikipedia.org/wiki/Program_evaluation_and_review_technique.

22 "Three-Point Estimation," Wikipedia.org, accessed June 8, 2021, en.wikipedia.org/wiki/Three-point_estimation.

23 Fred Lambert, "Tesla Is Collecting Insane Amount of Data from Its Full Self-Driving Test Fleet," Electrek, October 24, 2020, electrek.co/2020/10/24/tesla-collecting-insane-amount-data-full-self-driving-test-fleet.

24 Shivam Srivastava, Kartik Trehan, Dilip Wagle, and Jane Wong, "Developer Velocity: How Software Excellence Fuels Business Performance," McKinsey & Company, April 20. 2020, www.mckinsey.com/industries/technology-media-and-telecommunications/our-in-

sights/developer-velocity-how-software-excellence-fuels-business-per-formance?cid=eml-app#.

25 Charity Majors, "Observability—A 3-Year Retrospective," The New Stack, August 6, 2019, thenewstack.io/observability-a-3-year-retrospective.

26 Kelly Anne Smith, "Why Did the Robinhood App Go Down This Time?" Forbes, August 31, 2020, www.forbes.com/sites/advisor/2020/08/31/why-did-the-robinhood-app-go-down-this-time/?sh=b71543f6359f.

27 Niladri Choudhuri, "The State of DevOps Report 2019 Is Out," DevOps.Com, September 4, 2019, devops.com/the-state-of-devops-report-2019-is-out.

28 Chris Jones, John Wilkes, Niall Murphy, and Copy Smith, "Service Level Objectives," O'Reilly Media, Inc., 2017, sre.google/sre-book/service-level-objectives.

29 Hanno Ketterer and Christian Schmid, "Leaner, Faster, and Better with DevOps," BCG Global, March 22, 2017, www.bcg.com/publications/2017/technology-digital-leaner-faster-better-devops.

30 Sanjeev Kumar Punia, Anuj Pumar, Kuldeep Malik, "Software Development Risk Management Using OODA Loop," International Journal of Engineering Research and General Science Volume 2, Issue 6, October–November, 2014, ijergs.org/files/documents/SOFTWARE-135.pdf.

31 Stéphane Garelli, "Why You Will Probably Live Longer than Most Big Companies," International Institute for Management Development, December, 2016, www.imd.org/research-knowledge/articles/why-you-will-probably-live-longer-than-most-big-companies/.

32 Ilan Mochari, "Why Half of the S&P 500 Companies Will Be Replaced in the Next Decade," Inc., March 23, 2016, http://www.inc.com/ilan-mochari/innosight-sp-500-new-companies.html.

33 Mark Bertolini, David S. Duncan, and Andrew Waldeck, "Knowing When to Reinvent," Harvard Business Review, December 2015, hbr.org/2015/12/knowing-when-to-reinvent.